WELCOME

To our newest member,

Welcome to Sakura. It's great to have you with us. Thank you for picking up a copy of 'Sakura: Coincidence or Fate?' and being a part of this whole crazy escapade.

I have written this book because I went travelling to Lao and was truly inspired by what happened to me there. Not only does it have incredible natural beauty, it also has some of the friendliest people in the world and the energy of the place is amazing. The story I came across was quite different from anything I'd heard before and I felt it needed to be told.

This is that story. I hope you enjoy reading it as much as I enjoyed writing it.

If you want to find out more, be kept up to date with developments, want to visit Lao, have any feedback or want to get involved with any of the projects highlighted at the end of the book… or just want to say hello, you can contact me via SakuraRobson@gmail.com or the social media below.

For now, sit back, relax and enjoy the ride. Whether you're lazing in a hammock, on an exotic beach, on your favourite sofa, on a bus, plane or train… or wherever you may be on this fine planet of ours.

Enjoy and I'll check in with you later.

Robson
Moscow
Email: **SakuraRobson@gmail.com** Web: **www.RobsonDob.com**
Instagram/Twitter/Facebook: **@RobsonDob #SakuraBook**

P.S. If you are so inclined, why not post a photo on social media of you reading this book in some random location and tag @RobsonDob using the hashtag #SakuraBook. We've got a map with all the locations across the world where it has been read! Even before publication, we've chalked up London, Hong Kong, Bangkok and Las Vegas.

SAKURA: COINCIDENCE OR FATE?

Luang Prabang
Slow boat from Chiang Mai
SAKURA
Vientiane
Pakse
Don Det
120 km

SAKURA: COINCIDENCE OR FATE?

ROBSON DOB

A
ROCKET ANGEL
PRODUCTION

Published in 2018 by Rocket Angel Media

The names of characters and locations have been changed, as have certain physical characteristics and other descriptive details.

Some of the events and characters are also composites of several individual events or persons.

Rocket Angel Media

A CIP catalogue record of this book is available from the British Library.

ISBN: 978-1-9997966-1-7

CONTENTS

Coincidence /koʊɪnsɪdəns/

Noun. A remarkable concurrence of events or circumstances without apparent causal connection.

Fate /feɪt/

Noun. The development of events outside a person's control, in a way that cannot be prevented or changed, regarded as predetermined by some other power.

PROLOGUE

THE REVOLUTIONARY

'Steve Jobs, Mick Jagger, Ché Guevara… Stephen Sakura. They are all revolutionaries'

'Stephen Sakura?! He isn't a revolutionary.'

The first opinion is that of Alonso, a seasoned traveller from Brazil. The second is that of the man himself: Stephen Sakura. Sakura isn't his real name. I'm not even sure if Stephen is either. All I know is that he manages the Sakura Bar in Vang Vieng, Lao PDR. If you believe local backpacker myth, it is the most famous bar in Southeast Asia. That is certainly the view of Alonso. As a warm Saturday night draws to a close, Alonso sits in awe of this mystical man.

'You *are* a revolutionary! Do you know you have created the most famous bar in Southeast Asia? Man, I see your T-shirt everywhere… everywhere! First I saw it in Bali. Sakura Bar, Vang Vieng. What is this? Then I saw it in Kuala Lumpur, in Bangkok, The Philippines. Do you know it's everywhere?'

'That's what people keep telling me,' Stephen chuckles.

If you have travelled in Southeast Asia recently, you will have no doubt spotted people wearing brightly coloured vest tops sporting the Sakura flower logo. The name originates from the Japanese Cherry Blossom festival. 'Drink Triple, See Double, Act Single' is the infamous quote that adorns the back. Five hundred thousand of which have been given away over the years.

From Thailand to Vietnam, Indonesia to Myanmar, travellers know the name. They've been there, done that, got the T-shirt. Literally and figuratively. And they flock, in their hundreds, nightly, to this Laotian backwater. And here he stood. The man behind the bar behind the T-shirt.

He doesn't get caught up in all of the hype though. He remains as humble as ever. He has work to do. Eight years living in Lao and he still thinks he is only just getting started. And boy does he have some stories to tell.

When people ask me why I went back to Vang Vieng I tell them it's because I didn't have the blue Sakura top. I do it smiling wryly. They never know if I'm joking or not.

'That's a long way to go for a T-shirt,' they say.

The truth is I do have all of the other colours. Enough to create a rainbow in my backpack. The thing is, I didn't have the blue. I didn't have the white with pink trim either, but that's another story. My friend Armstrong has the blue. He'd probably even have given it to me when I got back home, but that just wasn't the point. I needed to go back. There were adventures awaiting me there and stories to tell.

They are right. It is a long way to go for a T-shirt, but there was so much more to it than that.

This is a story about many things, but in principal it is a story about a T-shirt.

PART ONE

LAO

CHAPTER ONE

THE T-SHIRT

The Sakura story started for me in Luang Prabang in Northern Lao PDR. I was there on a recommendation from my friend Lex who I had met in Vietnam five years before. He told me that Lao was his favourite country in the world. Not only was it number one, it was head and shoulders above anywhere else he had been. He should know, he makes a habit every now and again of selling everything and taking off. He goes around the world for 500 days at a time and calls it a 'Lexpedition'. He has been to 72 countries and Lao was his number one.

'Trust me, just go.'

'Why?'

'You will love it. I could say it is because of the people, the scenery, the beautiful temples. I might even say because it is so natural, unspoilt and laid back… but that doesn't do it any justice at all. You have to go and experience it for yourself. There is a special energy to the place and I think you will find yourself there.'

With a glowing recommendation like that, how could I not? Those eerily prophetic words of wisdom were still ringing in my ears when I booked my flights. I told my friend Armstrong and he was keen to join. From Hong Kong, but brought up in Britain, he is an artist with a tech start-up. Read as: flexible to leave the country at a moment's notice. We knew each other from our old working days as graduates at a marketing consultancy. That was ten years ago and we have stayed friends ever since. We had visited Goa together. That trip had inspired Armstrong to quit corporate life altogether. As a result, he is my go-to person for adventure holidays to obscure destinations.

For those who don't know it, Luang Prabang is a wonderful, unique and charming little town. It is famous for its beautiful and well-maintained temples as well as its slow pace of life. In 1995 the whole town received UNESCO world heritage status. Cornered into a long thin wedge where the Nam Khan river meets the Mekong, it is only three streets wide. It has an intimate feel to it with the French colonial influence. Fabulous little French bakeries line the main street. You can munch on a freshly baked baguette whilst drinking a Laotian coffee with condensed milk and watch the world go by.

Armstrong and I had just finished breakfast and were wandering down the main street. He turned to me and asked, 'What's this Sakura thing?'

I looked at him confused. I had no idea what he was talking about. He motioned towards a girl wearing a grey vest top. A black flower logo appeared on the front with the words 'Sakura Bar Vang Vieng'. The words on the back read 'Drink Triple, See Double, Act Single'.

I dismissed it in an instant. A cheap and tacky vest top. 'Some bar,' I shrugged at him. It didn't seem particularly riveting or unusual. I was also feeling a bit grumpy about clothes in general after the airline had failed to fly my bag out with me. Instead of going to Vientiane it had ended up in Vienna. Close but no… actually, not close at all.

Armstrong continued, 'It's just that it's the third person in a row I've seen wearing one since we left the bakery.' This intrigued me. Without even thinking about it I tapped the girl on the shoulder.

'Excuse me, what is this Sakura thing?'

The girl looked a little startled but then she smiled and motioned to her T-shirt. 'You've never heard of it? Sakura? It's a bar in Vang Vieng. You have to go. Everyone goes there, they give out free drinks and it is so much fun.'

'Vang Vieng?' I am pensive. 'But I thought there is nothing there now that the tubing has stopped.'

I'd heard of Vang Vieng. The infamous town made the headlines for all the wrong reasons in the past. The main activity there is tubing. You ride a blow-up ring down a river and get drunk and more along the way. Not sounding like the safest activity in the first place, 27 people had reportedly died doing it. We'd heard that the tubing had stopped in the interest of safety and that it didn't happen anymore. I felt torn as it sounded like it was quite some party! Although Armstrong and I would both like to think we were too old for those sorts of shenanigans, I had a sense that we had missed out on a unique experience.

The German girl frowned. 'No, it's beautiful… one of the most beautiful places I have ever been. It is a chilled out place and you can also party too.'

'What about the tubing?'

'They still have it. You can do it if you like.'

'Isn't it dangerous?'

'I think they have made it much safer now.'

'Did you do it?'

'Yeah, it's crazy. We had so much fun. It is like nothing you have ever done before. Are you going to go?'

'Unfortunately not. We have just booked flights to Bangkok for the day after tomorrow. Maybe another time,' I said, a little dismayed.

We chatted on the roadside to the girl and her friend who was also German. They had both been travelling independently and met in Lao's capital city, Vientiane, where they decided to join forces. One was heading on to Northern Thailand whilst the other was returning to Bangkok. I love this about travelling; the randomness of it all and the people you meet. You hang out for a while, travel for a time together and then you continue on your different journeys. Everything comes together for a moment, a single moment in time, and then life continues.

They had been in Luang Prabang for a few days now and seemed to share the common view: it is a beautiful place which is both chilled and quiet. They asked if we had been to Utopia. Zen by day, funky by night. We had. And it was both zen and funky during the allotted time periods that it said it would be. That was pretty much the only bar in town but it closes early. We asked if they'd been to the waterfall. They had. Bowling? Yes. Temples? Of course. That pretty much covered all the bases.

I had been travelling for the best part of two months at that stage in early July. I was on a 3-month sabbatical or #100dayweekend as I liked to call it. I had finished one job at the beginning of May and was due to start a new one in August. May was Friday, June was Saturday and July was Sunday.

August, boo-hiss, was Monday. I had a stag-do at the end of May and the corresponding wedding at the end of June, which meant that I had to keep returning to the UK at monthly intervals. I had made it work for me though. Having spent 'Friday' in Byron Bay, Australia and 'Saturday' in Hawai'i, I was now idling away my lazy 'Sunday afternoon' in Lao. In my early thirties, I was enjoying this new found freedom between corporate careers. Having struggled with the notion of turning thirty, this trip was exactly what I needed. I had a rebellious streak in me ever since hitting the three-decade marker. From being corporate and clean cut I had decided to grow out my hair long. I have big curls and a big beard to match. I am somewhere at the intersection between rock n roll, yoga and surfer. I am not quite sure how that's turned out!

Our original plan was to spend eight days in Luang Prabang and see where we ended up next. We had been up Mount Phousi to see the famous temple on the hill with incredible views across Northern Lao. There are also many temples along the main street that entice you in with gold pagodas. We were sure to take a Lonely Planet style photo with some monks by a stupa. We ate some good French food and sampled some of Lao's delicacies. We wiled away an afternoon or two in a coffee shop; Armstrong focusing on his business and me on my music.

When I was at school I had aspirations of being a DJ. I remember saving up my money for two years and buying a pair of Technics 1210s. That was a special day. I practised for several years and became pretty good at it. I had started playing out at clubs across Manchester and got my own radio show. I was also learning to make my own music. When I got a job, sadly, life took over. I found myself playing on them less and less as I didn't have so much free time on my hands.

Having some spare time between jobs, I wanted to revisit music production. Coming to Lao would help me do that. I had had lots of fun in my twenties and I was moving towards a healthier lifestyle now since I had turned thirty. I replaced disco dancing with downward dog. Asia would serve as a chance to do some yoga, meditate and refresh.

We were a little templed out and the slow pace of life was taking its toll. There are only so many croissants you can eat and coffee you can drink. We needed an outlet for our pent-up nervous energy. Luang Prabang had served its purpose.

I wanted to return to the Sanctuary on the island of Koh Phangan, in

Thailand, and do a week of yoga and meditation. I had been there before and it is one of my favourite places. People from all over the world go there to relax and detox. From hippies to backpackers, burnt out executives to celebrities. It is only accessible by boat, three bays up from Haad Rin beach. On its own secluded cove with huts all the way up the cliffs; it is blissful and peaceful.

I wanted to reconnect with nature, do yoga and meditation, detox and generally work on myself and my music. Armstrong was also keen so we had booked the flights to Bangkok. From there I was open to ideas. Perhaps go to Ibiza for the last two weeks of my hundred day weekend and get back to where the music all began. I had started to get an appetite for a party. Armstrong was ready to party too. I could tell from the way he was talking to some Dutch girls the previous night. Hell, maybe we'd do a night or two on Haad Rin's infamous full moon party beach on the way to the Sanctuary; it would make the detox all the more worthwhile.

On further reflection, it is also a hedonistic ying to the holy yogic yang. If you have done yoga intensely for a week you will know how much you desire a glass of wine or a cold beer by the end of it! Despite how holistic one might feel I say life is all about balance. Partying and detox are just an intrinsic part of that. We had it all planned out.

Or did we?

That evening Armstrong and I discussed our plan. Armstrong had to be back in the UK for what he called an 'urgent meeting' the following week and so couldn't extend his time in Asia. I later found out this 'urgent meeting' was, in fact, a date with a Norwegian girl he'd met in Goa. Of course, a girl. We're so easily swayed! Enlightenment on hold.

It became abundantly clear that we were not going to have enough time to enjoy the Thai islands. By the time we would arrive, he would have to leave again. At this point, I was feeling a sense of energy pulling me back to Lao and away from Thailand. I had a sudden feeling that my journey there was not complete, it felt like I had more to do. It was peculiar.

We agreed that we would cancel our trip to the islands but that left us in a quandary. What should we do instead for the next three days? We definitely didn't want to spend any more time in Luang Prabang. As beautiful as it was, and I would certainly recommend a visit, three days is enough.

It was at that point that we recalled the conversation with the German girls in the street.

'The one wearing the T-shirt from that bar, what was its name?'

'Uh, let me think, Sakura?'

'Ah yes, Sakura Bar. That was it!'

'They had recommended Vang Vieng in a big way.' It seemed perfect. It was a six-hour bus ride there and six hours back. Armstrong could make his original flight and I could come back with him and go to the islands. I even changed my Bangkok Airlines flight in order to go on the following Monday. As I recall this now, I have to laugh to myself because I had forgotten all about that. That was the original plan. It soon changed.

We could now visit Vang Vieng and make our own minds up for ourselves. We could chill out in the natural beauty of it, go to this bar, I could pick up one of those T-shirts — I could do with some more clothes — and experience what was left of tubing if we so wished. I think we were attracted to it because they told us how chilled and beautiful it was, not necessarily for any other reason. At that point, we decided to go to Vang Vieng.

CHAPTER TWO

BUS PURGATORY

I awoke the following morning from a restless sleep. I had a feeling transport around Lao would be difficult. I was worried that we would be limited by the number of travel options which would ruin our changed plan.

Whilst Armstrong slept, I went down to our guesthouse reception and asked about availability to Vang Vieng. It was about nine o'clock in the morning and the receptionist was only aware of one bus and that had already left at eight. I was annoyed at myself. Maybe this revised plan wasn't such a good idea after all and we'd be stuck in Luang Prabang for another couple of days.

He phoned the bus station to check and, lo and behold, there was a bus that left that afternoon at 14.30. It just so happened to have only two places left on it. Was this a coincidence? It would get into Vang Vieng, or 'Double V' as those in the know call it, at 20.30 in the evening. That sounded alright to me so I booked it and went to find Armstrong to tell him the news.

Things were working out and by nightfall we were going to be in Double V. Our tuk-tuk arrived at 14.00 to take us to the bus station via a number of guesthouses to pick up others. The first people to get on our tuk-tuk were a couple of young Welsh girls who were backpacking for the first time. Next to join us were a couple called Beef and Eliza. A Spanish girl named Pinar then joined. You could see that she had already taken

Beef's eye, which was odd, because we thought he was travelling with his girlfriend, Eliza. He was certainly engaging in some extracurricular flirting.

We arrived at the bus station and all I can tell you is that the searing mid-afternoon heat was overwhelming. Once we had finished with the cordialities; the hellos, the where-are-you-froms, the where-have-you-beens, everyone fell into the rhythm of the slow moving bus in silence. It was broken once every hour or so when Beef complained to the bus driver; asking for the air conditioning to be put up. He was duly ignored every time. I just focussed on not fainting and listening to my tunes.

The only thing that stopped me from completely passing out, or screaming, or getting them to stop the bus so I could walk back to Luang Prabang was the sheer beauty that I witnessed on that drive. As soon as you leave the city limits of Luang Prabang, the road bends around and up into the mountains where the landscape is spectacular. Lush green hills spread out in all directions. Completely unspoilt and with no sign of any human activity or development. The bus rode up one hill and down the next, from one valley to another.

Despite the heat of the afternoon sun, things did cool down and we stopped off for sunset at an incredibly beautiful spot for a ten-minute rest atop a hill. In the middle of the countryside, you certainly get a feel for just how scenic and natural Lao is. In the dusk haze looking out you could see many rolling hills as far as the eye could see. It felt so serene. So untouched. Like how the earth is meant to be. How it was originally, before civilisation came.

We all got back onto the bus and the monotony of the drive continued in silence. We had no idea how much further it was. After the sun went down, darkness set in. We were driving through such wilderness and there were no street lamps which meant you couldn't see a thing out of the window. The bus continued on for three more hours.

Suddenly, without warning, the bus pulled up and we were in Vang Vieng. It happened so quickly I didn't really know what to do. I hadn't quite thought past this never-ending bus journey actually ending. I was at a little bit of a loss. Wow, I had to get up, I had to find my hotel. I had to say goodbye to all of these new friends I had made. I wondered if any of them wanted to go for a drink perhaps. But equally aren't they all shattered and want to get on and find their way? I was desperate for a Beer Lao at this point and I knew Beef felt the same. I could feel it. I also felt the need to talk to him. I'd sensed an energy about him in the Luang Prabang bus station.

Beef's ears pricked up when I told Pinar I was from Spain, a coincidence or fate that we will deal with later. We all agreed that we would go and find our hotel and then meet back at the corner bar for a beer opposite where the bus had dropped us off. Armstrong and I jumped in a tuk-tuk, found our hotel and returned 15 minutes later. Vang Vieng is a small place and although we seemed a way away, we were close enough to the action. We rejoined Eliza, Beef and the Welsh girls we'd met on the bus at The Corner Cafe and we ordered some Beer Lao.

Wherever I go, I like to drink the local beers. On my first trip travelling I landed in Bangkok and that first Singha had tasted so good. The taste of pure freedom. As a result, whenever I have a Singha now I can still get that feeling. Over the years loads of people I have come across mention how much they love Beer Lao. It is the most prominent beer in Lao and it is actually an excellent brew. After that long bus journey, it tasted crisp and fresh. The perfect antidote to an eight-hour journey in stifling heat.

It was here, at The Corner Cafe, that we would find out Beef and Eliza were not a couple, just two friends travelling together. He was from Manchester, teaching English in Bangkok, and she was a New Zealander; a physiotherapist living in Vietnam. He was intrigued by my Spain connection. He had an ex-girlfriend who used to live in the same town, who it turned out I knew. She was a Russian and had gone to my school. As it happened I had dated her best friend, a Swedish girl. Would you believe it? Small world. Incredibly small world. That two people who randomly meet on a bus in Lao both dated a pair of best friends, a Swede and a Russian back in Spain.

We had some food and decided it was time to go to Sakura. We knew we had missed the free drinks but thought we would go there anyway seeing as that was the reason we were in Vang Vieng. We rocked up and it wasn't quite what I was expecting at all.

CHAPTER THREE

SAKURA

I hadn't really known what to expect but when the Germans had described the place as 'beautiful and chilled' I had assumed a bar with a view.

My first impression was that it was very basic and open. It was a wooden shack with no walls and a high ceiling, like a pyramid. The decor was timber everywhere. As we entered, there was a small outside space with a bar on the left illuminated in neon purple. This was a relaxed, gravelled area with a number of people standing around, drinking and mingling. There were a couple of beer pong tables set up and people were playing with excitement, in the throes of competition. As I went up a set of stairs I could look out across the dance floor in the middle of the 'pyramid' with wooden bench seating around the outside. The party tunes were pumping out and the atmosphere was electric.

There was a sea of about 200 people all wearing the T-shirt in as many bright colours as you can imagine. Neon pink, orange, blue, red, yellow, green and so on. Everyone was dancing and smiling. Groups of people from so many different countries integrated and chatted away to each other. Laughing, smiling, dancing. Instantly I could feel there was something very different about the vibe of this place. It wasn't like any other bar I had ever been to. Everyone just seemed so happy and friendly. Some played beer pong whilst others inhaled balloons of laughing gas. If someone bumped into you then they would apologise, offer you a high five and ask you where you were from. New found friendships were sprouting up in all corners of the bar. There was a raised podium in the middle of the dance floor packed full of people. On closer inspection, I later realised it was an old pool table. Not sure you could play a game on it now; it was packed with people on top raving away. There was a contingent of Koreans really going for it. Chanting with joy, hands in the air and jumping around. They certainly knew how to dance and they were going for it, screaming and gyrating to the beats. They really embodied the energy of Sakura and it was contagious.

At the back, beyond the dance floor, was a second bar with balloons being blown up and inhaled. These were laughing gas. 'Whooosh,' you could hear the balloons being blown up every few seconds with groups of travellers inhaling them.

Of course, everyone was wearing the T-shirt; the uniform of the hedonist. The badge of honour. They were being handed out with every order. Two vodka drinks was all it took to secure one. They had a selection of about five on display behind each bar. On that first night there was blue, orange, neon green, red and white. Armstrong went to the bar and ordered us two Vodka M-150s. I turned around and he was already sporting his blue Sakura top with pride. He was beaming. Just about as happy as everyone else in the joint.

It was at this point that I spotted Mr Sakura for the first time. A man behind the bar with a crazy haircut and a big handlebar moustache. Long, wavy, silver hair; shaved at the sides and tied up in a bun at the back. A silver fox he was. Perhaps he had dyed it silver. It was hard to tell his age in that light but Beef reckoned he was in his mid-thirties. He was definitely the master of ceremonies. He was orchestrating proceedings. The ringmaster to this circus.

He and his team had their hands on their hearts and were almost transfixed in a trance-like state. They were chanting in unison:

'As we say here in Vang Vieng;
Hats off heads,
Hands on hearts,
Praise thee father goose master shotgun up above,
May these beers go down smoothly,
Efficiently,
And with minimal spillage,
And absolutely no after sickness,

Three us!

Two us!

Gun us!'

Then they each downed a can of beer in an unconventional way - through a hole in the side. They cut a hole in the side of the can near the bottom and held it to their mouth. Then they opened the can from the top. This allows gravity and air to do it's thing, creating an efficient beer-delivery system. Down in one. I didn't really understand it.

Annoyingly, I was desperate for another drink and we were really ready to party having spent so much time on that bus from hell.

I went to buy a round but realised I had missed last orders. The chant and beer-downing signified closing time. They were now closing up the bar and Mr Sakura was on the microphone thanking everyone for coming and reminding them to 'Drink Triple, See Double, Act Single.'

I thought I would ask Mr Sakura where a good place was to party so I waved to get his attention. Most bar owners want to close up and get rid of the punters as quickly as possible. I can understand that. They want to go out and party themselves. He certainly looked like he wanted to go and party.

I said to him, 'Excuse me, we've just arrived in town and we were really excited to come to your bar but you're not serving anymore. Is there any place you'd recommend where we can get a drink?'

From my own experience of bar owners - and maybe you would say the same - they usually just ignore you, tell you to go away or generally try to get rid of you. Mr Sakura did none of these things. His response took me by surprise and changed the events of time as we know it.

He said to me, 'This is your first night in Vang Vieng? Welcome.' He said it with such a warmth and sincerity as if he really meant it. He said, 'You should go to Bar Fly. Take a right out of here, then take a left and it's about 200 metres down the road. Follow the crowds. It might take some time but everyone will end up there. That's where we are all going to end up. We look forward to seeing you here tomorrow. Free drinks on the house 8 'til 9.'

I thanked him and was sure we would return the following night. We followed his advice and went to Bar Fly where we partied hard and had a massive night. From Beer Lao, we moved on to the local moonshine vodka called Lao Lao. It had a nice chemically taste to it and we drank it with

M-150 which is the original Thai version of red bull. Only about 10 times as strong, mind, and the perfect fuel for some dancing.

I warmed up my dance moves by doing the Tae Kwon Do warm up routine I used to do some 10 years ago when I practised the martial art when I was young. Vang Vieng is packed with Koreans. Their national sport is Tae Kwon Do so it seemed apt. In fact, a few of them even joined in with the warm up.

Beef requested some music that never got played, a recurring theme transpiring when he asks for something. The Geordie DJ called Beef over and pointed at me dancing in the middle of the dance floor. I'd blended in some Mick Jagger to my Tae Kwon Do moves. 'Do you know that guy?' he asked, incredulously. 'He's a legend he is.' That is how Beef recounted it, I am not sure if he was just saying that to humour me.

My dance moves were definitely getting some attention. Especially from the Koreans who love to dance too. Even more crazily than me in fact! The party went on into the night. By the time we finished in the early hours I was quite hungry. We hadn't eaten much given the long bus journey.

Fortunately, the locals have caught on and there was an array of street vendors serving up various delights; from fried rice to chicken baguettes. Armstrong chose a baguette with everything on it. Literally. Chicken, bacon, ham and cheese. He then put as much sauce on it as possible, spraying it everywhere. I think the Lao Lao vodka had taken effect!

I personally went for a baguette with three Laughing Cow triangles. Laughing Cow? I know, I couldn't believe it either! For me, being able to eat a Laughing Cow baguette on the roadside at 3 o'clock in the morning was just what I needed. I have often been criticised for this choice as it was considered too plain and too dry. I disagree. I think 'Cow' gives the baguette the right amount of moisture and flavour. The beauty of this choice is that it's hard to get wrong. Given the synthetic nature of the cheese and the absence of salad or undercooked meat it's not going to give you an upset stomach either. Make sure you get three Laughing Cow triangles. They always want to scrimp on this and give you two. If you're still hungry afterwards have a second one like I did. Then you can proclaim to everyone that you have completed a full wheel. Always wheel when it comes to Laughing Cow!

CHAPTER FOUR

TUBING

Tubing.

Tubing has many a myth and legend associated with it. It is undoubtedly what put Vang Vieng and maybe even Lao PDR on the map, although, it is much debated as to whether this is a good thing or not. Regardless, the world over, if people have heard of Lao they will more often than not have heard of tubing.

In simple terms it is effectively a pub crawl down a river using the inner tube of a tractor wheel, like a big rubber ring, to float down between each bar. If you search Vang Vieng on Google one of the first things that comes up is an article from the Guardian newspaper from some years ago. It describes how young backpackers get drunk and out of control along the river and have been known to die doing it. It was therefore labelled a dangerous activity that you should avoid.

It got so bad that apparently there was a big governmental clampdown after 27 people died in a single year. This put an end to this horror show and blight on Lao's international reputation. Those in the know will tell you it still goes on but in a more organised, safer and controlled way.

We'd heard and read all of the above before we arrived in Lao PDR and as I had intimated earlier, we had no plans of visiting Vang Vieng let alone go tubing. Whilst our interest had been piqued by the German girls telling us that it still went ahead, we were not sure if we would try it. It seemed reckless and the trip wasn't really about getting drunk. I have to say I secretly wanted to do it though! Seeing as we'd come this far it seemed silly to miss out.

We were hungover that day following our arrival in Double V and after all the fun of Bar Fly. I opened my eyes to feel the throbbing pain in my head. It was so bad I just wanted to rip it off and throw it on the floor to get rid of the pain.

This soon changed when I opened up our curtains. Because we had arrived at our hotel in the dark we hadn't seen what was out there, but in the morning light the view took my breath away. Our bungalow backed on to a garden and then the Nam Song river. Beyond the shore arose these large limestone karsts. Probably several hundred metres high. Rich in vegetation and greenery. The orange rock glinted in the morning sunlight, varying the colour of the karsts depending on how the sun hit it. Pure nature and beautiful. These rock formations just outcropping out of nowhere towering over the valley where the river bed was. It was enough to help me start to feel better about the hangover already.

We satiated our hunger with breakfast at a little cafe where we lounged on triangle cushions and watched Friends. For some reason Friends reruns are on back to back all day long in almost all the cafes in Double V. To be honest, it is the perfect antidote to any hangover. Sitting relaxed, lying on a triangle shaped cushion, having some food and a coffee, and not having to concentrate too much on the show or anything for that matter.

We finally mustered enough strength to get back to the hotel to have a nap and that's when I got the call from Beef.

'How are you boys? Are we ready to try this tubing malarkey or what, Big Dog?'

Armstrong looked like death but that wasn't going to stop me. Bizarrely I felt short changed by my 3 a.m. finish. I felt that I had more to offer VV's party scene and I was getting in the mood for more dancing. I was going and so was Armstrong. He just needed a twenty-minute power nap. I could tell he didn't want to miss out.

'We came all this way, it would be rude not to,' I responded.

We had to get up there by four, otherwise, they wouldn't take us on so we got ourselves ready. We met up with Beef and Eliza and we were all set to go. We had to wait for a friend of Beef's called Rose. She was Scottish and in her sixties. Beef had met her at their guesthouse and asked if she wanted to come tubing with us. I was a little baffled because I wondered if he fully knew what tubing was all about and if he had explained it properly to Rose. When we met her, we found out that she was nothing short of a legend. She had been travelling around Southeast Asia on her own and had been doing a sterling job of it. We went down to the tubing sales office on a tuk-tuk. We got our tubes and we got our numbers. When you go tubing you get a number on your wrist. I got 142 in bold blue magic marker.

We were all set! Two chaps jumped into the tuk-tuk with us. One was half Peruvian, half Italian; Marco, The Latino Lothario. The other was an Israeli-Iraqi chap named Johnson. Well, it was Elo, Eli, or something, but he said his friends call him Johnson. They had met on the overnight bus from somewhere and had become instant friends. That's what I love about travelling. Just like the German girls. Marco had said he was supposed to meet up with some of his Brazilian friends somewhere on the tubing route. There were fifteen of them. I feared the worst. A load of muscle heads from the Copacabana, strutting around arrogantly could have the potential to spoil any vibe.

We got out and got on the river. There were no instructions or anyone there telling us what to do so we just did it. We jumped in our tubes, taking great care not to impale one's leg on the sharp piping that sticks out from the inner part of the ring. This was definitely an authentic inner tube. I would love to say we just jumped in the river on our tubes, with style and grace, but it just wasn't quite like that.

It was awesome on so many levels. The best level being that it was mid-afternoon and whilst everyone else back home would have been at work we were lounging on rubber rings floating down a river. The scenery was magnificent and there we were relaxing at nature's pace down the river.

It was at this point that I felt a real appreciation for just how spectacular the landscape is in the area surrounding Double V. The Nam Song river runs through a valley with these limestone karsts protruding up from the banks of the river. Straight up, looking at these huge rock formations from the comfort of my tube. The orange colour had changed since this morning as the afternoon light now hit them. Up close, I could now see all the trees teeming with wildlife.

The tubing river was a lot thinner than I thought it would be and also a lot shallower. The rate at which you move is very slow and is quite nicely paced. Chilled. You just amble down the river.

We had lots of banter just shouting at each other and chatting. Occasionally we would link up with one another and go down in a sort of tube train and other times we just floated on our own. It was brilliant. We had a good laugh splashing each other and getting caught up on the rocks. At one point Eliza got caught up in a bush and was stuck. This was highly amusing until I managed to do the exact same thing a moment later.

The first bar we came across was on a hill by the river bank. We could hear the beat of dance music filling the air as we got closer. We saw people on the hill and caught our first glimpse of the tubing party scene. Quite a contrast from the peacefulness of just floating along the river.

Beef asked, 'Shall we stop and see what's going on?' I could only think of one answer to that question.

'Yes! Let's see what it has to offer.' There were some young local boys throwing rope out to us. I caught it and reeled myself in. Getting out of the tube was harder than you think it's going to be, and I did well not to slip on the rocks by the shore. I then put my tube amongst a big pile of other tubes that all look the same.

I commented to Eliza, 'We definitely aren't going leave with the same one we came in on because they all look the same. Aim to get one that doesn't have a spiky valve.'

We climbed up the hill in bare feet. We made it to the top and arrived at this bar, well, shack with a dance floor. There was a small wooden bar and a decking where everyone was hanging around. Beyond that was the greenery of the Lao countryside. The limestone karsts jutted up on the horizon. The setting was idyllic.

The bar was raving. Four in the afternoon. Raving. The music was banging, people were drinking beers, singing and dancing. There were a group of guys shotgunning some beers - this I had found out was the ritual we had witnessed the previous night where you cut a hole in a can of beer and drink it down. The whole thing was quite primal.

It reminded me of the days of Goa Trance. There used to be a scene there of parties on the beach in the early nineties. As an aspiring DJ I had wanted to go clubbing on a beach in those days but I was too young and missed that scene by a few years. I always imagined how awesome that would be. And here it was - different location, on the banks of a river rather than a beach, but an idyllic setting nonetheless. What an amazing party. A small number of like-minded people from all over the world coming together to have a rave in this natural splendour in the middle of the afternoon.

I bumped into some lads from Birmingham who absolutely loved my hair. I guess it is quite unique and distinguished. These big curls. Like a lion's mane. Anyhow, they loved it. 'Cracking lid you've got there,' one of them said motioning to my hair. That broke the ice. I was a little intimidated by them at the beginning to be honest but they were friendly enough. That was the beauty of it - people you would have crossed the street to avoid back home were your best friends here. Everyone just having a good time brought together by a sense of adventure and free spirit.

I did a shotgun of beer with them. My first ever. It was a struggle as well, I have never been one to down beer and given my hangover it was hard work. It did the trick though and I was certainly feeling spritely and energised after that. Armstrong had a water like a true party animal. Beef had a beer which he sipped and Eliza had a Diet Coke. Rose at least ordered a gin and tonic and received whatever the Lao equivalent was. Then we raved it up. Rose leading the charge to the dance floor and we all followed suit. Everyone was dancing, smiling and having a good time.

Marco, our Peruvian tuk-tuk buddy, found his friends, the Brazilians. It turned out to be one not so strapping chap named Alonso, who was actually a really nice guy, and his 15 Brazilian lady friends. Bikini clad and showing us all their dance moves. With the tunes energising us, it was time to get involved and, well, standing in a field dancing amongst them, was no hardship at all. Girls in bikinis, boys in board shorts; something for everyone. All the beautiful people. Smiling and raving. No inhibitions. Everyone in a good mood and the tunes playing out in the sunshine.

We hung around there for a while and decided to check out the next tubing bar. 'If this is the first one, imagine how good the next one is going to be?' We got back on our tubes and onto the river. A little bit of a buzz, we felt good about ourselves. Party animals. Surrounded by party people.

The bar had been good, I'd had a beer. That set me up quite nicely and I was starting to feel it again. We just carried on down the river. The next bar we stopped at was aptly called the Final Bar which took us by surprise as we missed out a couple without even realising it.

There was no one in this bar when we got there. We thought perhaps we had made an error when we realised the party was in effect at that first bar. We knew people would end up coming down the river and come and join us eventually, so that wasn't a problem. In the meantime, we had free reign of the bar. Armstrong mixed the cocktails and I chose the tunes. We also ended up having a free shot of Lao Lao with every drink as well. It tasted disgusting but everyone had to do it. When in Lao, drink Lao Lao. Looks like vodka tastes like tip-ex. That I think helped start to perk people up, certainly those that could hold it down. With every drink, we received a multicoloured friendship bracelet in bright colours. I collected a few. Pink and green, red and yellow, orange and blue.

I remember Armstrong coming up to me at one point and saying he was going to have a Happy Shake. This was a remarkable turnaround for someone who was drinking water in the first bar. I asked him if he really thought that was going to be a good idea.

We had heard different things given we didn't know how we were getting home or how long the tubing route was. We knew we were in Final Bar but what did that mean? We didn't know how long the river was back to town, how long it would take or how many bars there were. We had heard it was another two hours into town which didn't make sense if this was the final bar. And you don't want to be tripping off your head if you've got another two hours of river to negotiate. Not in the dark anyway that was for sure.

We played pétanque and volleyball on the makeshift court that had been set up. As more and more tubers arrived they joined us. One by one adding to each of the teams on the volleyball court. There were some Swedes, a Dutch guy, a group of French, a Canadian and so on. Beef and I were chatting to a Kuwaiti and his Turkish wife. We drank cocktails from buckets that tasted of lemon sorbet. Really delicious and refreshing in the afternoon heat. We had the best time and met so many people. It was quite small, maybe one hundred or one hundred and fifty people. Well given my tubing number probably 141 other people to be exact. Quite small - that is what made it so good. Everyone was so friendly, everyone was really chatty, there were so many different countries represented. We danced as the sun went down.

The time came to move on. As it was now dark we were corralled to where many tuk-tuks were waiting to take us back to Vang Vieng. We went straight to Sakura to start collecting T-shirts in a big way. Which means drinking vodka drinks because it is: 'Buy two vodka drinks, get a free T-shirt'. Pink. I got the pink one first so I could be Mr Pink. It matched my pink tubing shorts. Double V is quite luminous. I then got a neon green Sakura top. An orange one too. That night, that was a good haul of T-shirts. They didn't have blue ones. They had sold out of those the night before. No blue.

I was definitely dancing on the pool table in Sakura Bar. Right in the centre of the dance floor is the pool table. I just got on because everyone else was on there. There were these two Korean guys, who were ripped and dancing with their tops off alongside me. They had the moves and their very own fan club of screaming Korean ladies. Amongst whom, Rose was dancing and screaming with. They were precisely the opposite of me, as in I am not ripped, am not the most coordinated dancer and my top was still intact.

I motioned that I was going to take it off and they along with their fan club encouraged me to. I thought what the hell, so I took it off and swung it above my head. The Koreans loved this. I then threw it in the general direction of where Armstrong and Eliza were standing, hoping they would pick it up. I didn't want to lose my pink Sakura top after I had just acquired it. They actually did catch it and I got it back the following morning which was a pleasant surprise. I would never have done this sort of thing anywhere else in the world but in Vang Vieng and especially in Sakura, there is such a vibe that there is no self-consciousness. No judging. No attitude. Everyone just wants to have a good time with one another.

There was a moment when Eliza tapped me on the shoulder and said, 'Rose wants you to take her dancing.' I took her by the hand and led her onto the dance floor AKA the pool table. She showed off her moves and she had a boogie with the Koreans. I tell you the Koreans, they know how to party. All my worry about whether Rose would have a good time or not had long since dissipated. She was a party animal. She got into it more than anybody else. She loved the tubing, probably even more than we did. She also ended up taking part in all of the tubing activities and was up for anything. She was the star of the show.

That was quite the night. We definitely had a great time in Sakura. We met more people, more friends and more nationalities. Everyone that we had seen on the tubing route was now in Sakura. You recognised them and if you hadn't talked to them before you talked to them now. The place was rocking.

When Sakura closed at midnight we got onto a tuk-tuk in search of what we heard was a Jungle Party. We didn't even know if it existed or not. Eventually, the tuk-tuk got moving and bizarrely it did actually exist. A nightclub in the middle of the jungle. We got a smiley face drawn on our stomachs. The tunes were good and it was decked out with fairy lights on the trees.

I met a journalist, Sally, who worked for the Guardian. I was quite excited by this as that particular newspaper had run the article on Vang Vieng and Tubing. Having experienced it that day I had seen it for myself and was making up my own mind about it. I quizzed her about what she and her paper thought. To be honest, I can't remember her answer. I can't really remember much from the Jungle Party, only that it was an awesome party to conclude an awesome day. The whole day had been perfect.

CHAPTER FIVE

THE IDEA

Unsurprisingly, the day after tubing, the day after two nights on the sauce in Double V, we felt a little bit worse for wear. Armstrong doesn't handle hangovers at the best of times. In fact, he is one of the worst people for hangovers you've ever met in your life. This is maybe why he is one of the best people in the world for partying. There seems to be a correlation.

I got a message from Beef saying he and Eliza were ready to move on.

'Do you know the meaning of Vang Vieng, Big Dog?' he asked, 'It means: "float on". We are going to do just that to the 24,000 islands.'

'Do you know the meaning of the word Lao?' I responded. I had read it in one of the guidebooks. 'It means: "don't rush".' I was staying.

I agreed that we would meet them for a final breakfast to say goodbye. Armstrong wasn't happy about this. It wasn't personal, he didn't want to see anyone at all at that point. Like I said, he does not do hangovers well. He just wanted to curl up into a hungover ball, eat as many carbs as he could and watch Friends episodes back to back.

We ended up having a really fun brunch. It was like the day after a wedding where everyone is really hungover. We were around a lot of people that we knew, who we had just recently met through the shared experience of tubing and Sakura Bar. Everyone is, of course, wearing a Sakura T-shirt. It is only a town of three streets so you are constantly seeing people you know; in cafes, wandering the streets, getting their buses and so on. People you'd had such a grand old time with the night before.

Beef and Eliza departed. Armstrong and I were left at Corner Cafe. Vang Vieng's own version of Central Perk. Amazing that only 36 hours prior we had arrived there and didn't know what was to happen. Yet now we had met so many people. So many new friends and had such a good time. A single moment in time. It hadn't happened before. It won't happen again. Ichigo ichie.

We had a carbolicious breakfast and decided to go kayaking for some ridiculous reason. We didn't want to waste the day and we also thought the physical activity would be good for us. Armstrong did not want to go tubing, that was for sure. I don't blame him, I didn't want to either.

The kayaking was good fun. We saw the river in a whole different light. We did the extended two hours of river that no one does on the tube. It was so peaceful and relaxing. I got a real appreciation for the whole river and getting new views of the limestone karsts as the river meanders around them. We saw the action at the bars as we went past and we both looked at each other. I could tell we were both very tempted. I would highly recommend the kayaking. It was an awesome experience in VV. The river and the scenery were stunning. We passed people in tubes and had a little chat to them. Germans, Koreans, Thais. We splashed them with our oars and splashed each other. Our guide was a brilliant Lao chap who was very mischievous and sprayed everyone with water. We also witnessed what life is like along the river for local Lao people. Everyone just very chilled.

We got back to the hotel and felt energised from all of the physical exertion. You could tell we were both thinking about it.

'What shall we do now?'

'Let's go tubing.'

'Why not, what else are we going to do?' We were only going to find a bar anyway and we were probably both ready for a drink.

We played Swedish House Mafia - Greyhound and David Guetta - Titanium. The combination of these two tunes together got us pumped up the previous day. They got us pumped up again that day too. We jumped around dancing on our balcony for those two songs and that was the tonic we needed to get ourselves going. We felt like we were Titanium.

Much fun was had. We went on the tubing route and this time a different set of bars were open. One in an old converted house which was quite cool. It felt like someone's house party… but imagine the best house party in the world.

We bumped into people we had seen the night before. We met up with various new people and, of course, went to Sakura again. We drank vodka and picked up some more tops for the collection. Another couple of green ones, a yellow and a burgundy. Still no blue, unfortunately. The party continued into the night, dancing on the pool table as standard.

We went to the afterparty at Lock Bar this time. The crowd was rocking with a good DJ on. The Koreans were going absolutely mental, jumping up and down and screaming. I really connected with the Koreans that night and in Lock Bar they were all going crazy. I enjoyed dancing with them. I even picked up a few new moves from them for my repertoire. I saw my friends on the podium again, the two ripped Korean chaps so we danced around some more and I managed to get my very own fan club of screaming Korean ladies. It was awesome.

It was so hot in Lock Bar that I needed to take a break. I was outside chatting to the Brazilian backpacker called Alonso. He was telling me all about how different the culture is in Brazil.

At that point, Mr Sakura came outside onto the terrace. He needed a rest as well. He said how tired he was. 'It has been a tough, long season and I am beat.' When I started to chat to him Alonso made the connection of who he was and was beside himself. He kept asking Mr Sakura if he realised what a revolutionary he was.

'You know it is the most famous bar in Asia?! You have the most famous bar in Asia. Everywhere, I see these T-shirts everywhere. Do you know your T-shirts are everywhere?'

Mr Sakura was smiling, 'Apparently so, I hear that a lot.'

I asked him how long he had been here. He said eight years and told me a couple of stories of his time in Lao. At that moment I realised his story needed to be written. I thought, he must have lots of similar stories. Everyone loves a good ex-pat story, breaking free from the shackles of western civilisation. Plus the fact that he has created the most famous bar in Asia. The actual bar itself is fantastic and overall it is the greatest party in the world. He is a really funny guy who is actually incredibly humble and has a crazy moustache and haircut. These things combined were the ingredients for a great story just waiting to be written.

I told Sally, my journalist friend, from the Guardian, 'His biography would be fascinating. You're a writer, you should really write it!' I was quite animated and excited about it.

'Nice idea, but I am busy travelling for the next year. Why don't you write it.'

That was the moment.

I had never written a book before but I felt inspired. So much so, that I found myself at 4 a.m. with Armstrong fast asleep in the next bed typing away the prologue on my phone. I woke up the next day and found it in the notes section. I had forgotten all about it if I am honest. I thought it was a good idea but then thought I probably wouldn't ever get round to doing it.

That morning Armstrong had to head off to Luang Prabang to catch his flight and we had one final breakfast together on the banks of the Nam Song at our hotel. Both feeling utterly broken but oh so happy; I can't even remember a time when I felt that level of happiness. We had partied for three days straight and we had had the most fabulous time. Just really happy about Lao and life. What a great place it was. Armstrong was genuinely sad to be leaving. I was supposed to go back with him as you may recall, because I had a flight to Bangkok on the Monday. I was having such a good time in Vang Vieng that I didn't want to leave.

I had to reconsider what I was going to do next. I didn't really know. I was thinking about joining Beef and Eliza in Don Det, the 24,000 islands, because they told me how wonderful it was there too. I had also had three days of partying and didn't want to leave just yet. In fact, I didn't even want to move and the thought of a 24-hour bus journey was a non-starter. I actually did very little that day. I went out for dinner with Sally, her sister Claire and their dorm buddies from their hostel. There were some Dutch girls and an Australian named Baz.

We watched the Wimbledon final at the Irish Bar and went to Sakura for a drink. Most of us were broken. Generally, the party was going on but I think at that point everyone I knew was partied out. Myself included. Mr Sakura was there, of course, on the microphone as always.

'Welcome to Sakura Bar. We have got happy hour going until 9 p.m. Remember folks, drink triple, see double, act single.' The man never stops.

I had a chat to him. It was the eve of his 32nd birthday. I told him that I was going to write a book about him. He didn't seem to think there was much of a problem with that. He seemed pretty happy with the concept. I showed him the prologue that I had written and he loved it. It was pretty easy to get his buy-in. That was cool.

Being his birthday the next day he was in a reflective mood. In that conversation he told me how he felt a little bit older now. He said he was coming of age. He was a party animal and now he probably needed to calm down. He also had great aspirations for Vang Vieng as a town.

He said that the sprouting of growth was happening and a fledgeling ex-pat community was coming together. He introduced me to three Italians who were setting up a Pizzeria - Il Tavolo. He was promoting them behind the bar by showcasing their pizza boxes. He said they had been busy building a proper pizza oven.

'The pizza there will be the best in town, made by real Italians with a real wood fired oven.'

They told me I should come back in two weeks for the grand opening. I said, 'Perhaps I should come back, write the book and try the pizza.' We all laughed. I didn't know if I was being serious or not.

The following day was Mr Sakura's birthday and the way he says, 'It's ma birth-day' is somewhat of a catchphrase. He does it whilst doing a little dancing jig.

'It's ma birth-day… it's ma mother-fucking birth-day!'

At 12 midnight he did his first traditional birthday shotgun. The tradition being that you do one for every year of your life during your birthday. That meant he had to do 32 shotguns in 24 hours. That equated to one every forty minutes. He had a tally count on his arm to mark each one off with a magic marker. At twelve midnight he had his first and clocked up the first notch on his tally. Several of his staff did it with him. I joined in too. [Are you playing along at home? If so, grab yourself a can of Beer Lao or equivalent and join us in wishing Mr S a Happy Birthday!]

'As we say here in Vang Vieng,
Hats off heads,
Hands on hearts,
Praise thee father goose,
Master shotgunner up above!
May these beers go down,
Smoothly,
Efficiently,
With minimal spillage,
And absolutely no after sickness,
Three-us,

Two-us,

Gun-us!'

Being Mr Sakura's birthday I had to stay one more night to celebrate with him. When we walked into the bar he was doing well all things considered. His shotgun tally was up to 22 and I was there to do number 23 with him. I did number 1 and number 23. I am glad I didn't have to do all 32. I gave him his birthday present. It was a yoga manual for prisoners and castaways by my teacher back home.

I have been doing yoga for 7 years with a very special teacher. She is my instructor and is one of the most inspirational people I know. Swami Pragyamurti Saraswati has been Director of the Satyananda Yoga Centre for over forty years. Her interest in the spiritual path began in the sixties when she came into contact with the teachings of the Bihar School of Yoga and its founder, Swami Satyananda Saraswati. On Swamiji's instructions she started teaching yoga in 1971 at her house, which we now know as the Satyananda Yoga Centre.

Over the last twenty years Swami Pragyamurti's main areas of work have included the establishment and teaching of a two-year Yoga Teacher Training Course, working with people living with HIV/AIDS, and teaching yoga in prison where she feels strongly about the importance and relevance of yoga and meditation to the inmates. Her warmth and wisdom demonstrate how the Indian tradition of Sannyasa can be integrated successfully with life in a busy Western city.

She is so inspiring and has led such an unbelievable life and I have found her teachings of yoga have really changed my life.

She has made a yoga manual specifically for 'Prisoners and Castaways' teaching you some yoga techniques when you are on your own and without access to a class. The funds from the sales of this allow copies to be given to prisoners to help rehabilitate them and find a better path.

I had been carrying this manual around with me my entire trip to do some yoga whilst on my own. I had been using it heavily in Byron Bay, Hawaii and now in Lao. I wrapped it up and gave it to Mr Sakura. He was quite touched and he really liked it. Especially seeing as he had been a prisoner and he was a castaway himself. I talked to a few of his bar staff and they thought it was really appropriate.

He told me that he was going to drop it off at his house so that he didn't lose it over the course of the night. I was intrigued as to where he lived, so he told me to follow him. I expected a walk to the outer reaches of VV and some palace on a hill. Instead we literally went across the street to a little guesthouse. The guy has the shortest commute in the world. He lives in one room in a guesthouse. No kitchen, just a microwave. It is very humble. A room with enough space for a bed and a bathroom. It wasn't the tidiest I will have to admit. A man of few possessions but a number of Sakura and Happy Bar T-shirts sprawled across the floor.

'Don't mind the smell,' he said as I walked in, 'That's just opium.'

We came up with a 'Terms of Reference' for the book. I think he was on shotgun 27 by this point. I thought he was pretty drunk so I was never actually sure if he wanted me to write it or not. Whatever state he was in though, I got his agreement. The terms of reference read like this:

TERMS OF REFERENCE

There is going to be a book.
This will, of course, be made into a Hollywood blockbuster.
Mr Sakura would be played by Johnny Depp.
I would be played by either myself or Brad Pitt*.

The asterisk denoted that if I didn't play myself then Brad Pitt would play me. He raised the point that I would be too busy opening Sakura Seoul and may not, therefore, have the time or the inclination to act. Equally, as I was not a professional actor, I wouldn't want to make lots of silly errors through not knowing the industry and lingo. I didn't want to piss Johnny off by slowing down the production.

I got his agreement there was going to be a book and that I would come back in a couple of weeks to do interviews. I got his agreement there was going to be a film, although, I am not sure that is within his control. I asked him for a white with pink trim T-shirt and that is where the agreement stopped. He said he couldn't.

'Oh that one? Yeah. It was a limited edition and there were only a small number printed for Canada day.'

He showed me a Happy Bar T-shirt from St Patrick's Day which he assured me was very rare.

I said there was one more thing. 'If I come back and write this book can I also do a DJ set in the bar?'

He said, 'If you write this book, I am sure we can organise you a DJ set. I think I can make that happen.' He chuckled, 'I know the guy who runs the place.'

We both laughed.

CHAPTER SIX

LEAVING VANG VIENG

The next day I woke up and all my friends had gone. I felt the need to go to Don Det and I wanted to see Beef, Eliza and the 24,000 islands. They were off to Cambodia after that and asked me to join. The decision was taken out of my hands somewhat based on my conversation with the ambassador at the British Embassy.

'Hi, I am a British national in Vang Vieng and I want to go to Cambodia. The problem is I have a full passport with no space for the visa…' This was annoying but I do have to say I count this as one of life's big achievements. I love to travel and this is my trophy to show it.

The ambassador informed me that I could get a special travel document that would allow me to go to Cambodia. The problem was I would have to wait until Friday for an appointment. I had no idea what day it was. He told me it was Tuesday. The document would allow me to travel to five countries but I had to state dates, and ideally, provide travel and hotel receipts.

Given the sporadic nature of my travel plans, this was going to be impossible. Beef and Eliza would have left by Friday so I would have missed out on the 24,000 islands. I wasn't even sure I wanted to go to Cambodia anyway. I talked to a travel agent and he said that on the Cambodian border they just add a page to your passport anyway should I want to go.

I thought, 'Screw it, I'll just jump on a bus and go to the 24,000 islands now and deal with it when I get there.' If I decided to go on to Cambodia with Beef and Eliza I would chance my arm at the border. And so began a 24-hour bus journey. One for every thousand islands I was going to see. What mega journey that was.

It started with a minibus to Vientiane, the capital of Lao. In the main bus station waiting for my next bus, I bumped into a couple from France who enquired about my T-Shirt; a Yellow Sakura top.

'Oh Sakura Bar? You haven't heard of it? It's in Vang Vieng. It's the best bar in the world. The place is really beautiful and chilled, and then the bar, it is really cool. They serve free drinks and if you buy two vodkas you get a T-shirt…'

I suddenly realised I was giving the same pitch the German girls had given Armstrong and I only a week previously. As I was saying it, even describing it just as they had done, it sounded good but it still didn't do the place justice.

'Just go,' I said, 'See it for yourself. It's amazing and you won't be disappointed.'

I could see them look at each other. Their minds racing.

'Yeah, it sounds great. We should go.'

'Yeah, I think we will. I mean we've seen the T-shirts everywhere!'

'Yes, EVERYWHERE!'

I left them contemplating how they could rearrange their travel plans and headed to the ticket counter to exchange my receipt for a bus ticket. Once I had my ticket the man at the counter said that the bus was leaving so I needed to hurry. I ran as fast as I could across the bus station to the spot where the bus was waiting. When I got there I found an empty bus with locked doors. A rabble of people hanging around waiting to get on. I recalled the meaning of Lao and smiled ironically to myself, 'Don't rush.'

'Don't worry, it isn't going anywhere soon.' He was wearing an orange Sakura top and she was wearing baggy elephant trousers like travellers do. I stopped and chatted with them. She was saying how happy she was that she could wear these trousers all the time here. Back in the UK attire like this was limited to cold Sundays at home.

'You don't want to leave the house, let alone cross continents wearing them.'

Her boyfriend was quite possibly the youngest person I had ever seen. Ever. They were cool. Really cool. He was from Newcastle so I gave him the nickname Geordie Sam. She was from the South but lived in Newcastle so I called her Honorary Geordie Liz or HGL for short.

We got the sleeper bus to Pakse. The sleeper buses in Lao are quite unlike any other bus I have seen in the world. They are big double deckers but instead of seats, they have little boxes about the size of a single bed and about half the length. These are classed as two berths for two people. Fine if you are a small person, not fine when you are 6'3" and sharing with a big German guy who is also over six foot. I noticed one of the other bunks was free and encouraged him to move over there.

This was a bumpy affair for 12 odd hours arriving into the small Southern city of Pakse at dawn. We changed onto a mini bus to take us further South to the banks of the Mekong. Finally, we took a boat from the riverbank to Don Det, one of the 24,000 islands. We played a game of arrival time sweepstake. Everyone on the bus guesses a fifteen-minute window of when they think the journey will end. Everyone puts 10,000 kip (USD 1) into a pot and the winner takes all! A great way of making friends with your travel companions. Use the proceeds to buy them all a Beer Lao when you finally get there.

By the time we got on the little boat to take us to Don Det it was midday the following day. We had been travelling 24 hours non-stop at this point. The sun was high and the sky was hazy. The humidity was oppressive. We all had that lucid tiredness that a lack of sleep brings but balanced with the wide-eyed adrenaline of being close to our destination. We could see Don Det from the banks of the river and knew we'd be on it soon.

It is a fascinating part of the world because the Mekong is very wide at that point. Wide enough to fit in 24,000 islands. It was so unique. It was a river, obviously, and so it is flowing in one direction at quite some speed. There are all these little islands, some are tiny and some are several kilometres long. They are built up of mud and vegetation. Loads of them stretching out in all directions.

The Mekong runs fairly rapidly and was brown and murky. I later found out it is actually very clean there and the brown colour is just from the silt because it was rainy season and the river was running wild. In rainy season the Mekong is brown and the land is green. In dry season the colours are reversed.

This small unsteady boat with an outboard motor took us to Don Det.

Negotiating through these little islands. The sun was high and hot. After all this travel we sensed a real feeling of achievement. Looking around at these tired souls everyone was smiling with curiosity. What is this place? And what adventures were awaiting us here on Don Det?

CHAPTER SEVEN

REINVENTING THE WHEEL

We landed on the small island of Don Det after this epic 24 hours of travel. First impressions were that it was very hot, humid and laid back. It is a two kilometre long silt island. Vegetation has grown on it giving it a rainforest landscape. There are no roads, just thin mud tracks leading all over the island. Being a tropical climate there were plenty of puddles to splash about in. The infrastructure is all made out of bamboo and there was something very Robinson Crusoe about it.

I didn't have a guesthouse and didn't know where I was going. I had no idea where Beef and Eliza were staying either. The Geordies had already booked into the only place on the island that had a swimming pool which seemed like a wise choice, so I went with them. We then found out it was the best part of a 40-minute walk to get there and we took the long way round as well. It was a scenic walk through jungle and rice paddies, although, I just would have preferred to do it without backpacks in the midday heat.

When we got there it was empty and I personally thought it was not where I wanted to stay. I got chatting to the English guy who was helping out there. It seemed to be the case in Don Det, I found, that there would be random foreigners fronting up establishments like hotels and restaurants. He recommended I go back to Sunset beach rather than where we were which was Sunrise beach. He suggested a few guesthouse options.

I left the Geordies and we exchanged numbers. I headed off on the quicker route back to the top of the island. I had done an entire revolution of the island before I had even dropped my stuff off and was kicking myself. Although it did mean that I managed to get my bearings very quickly. It also meant the following would not have happened. A coincidence or fate perhaps?

On my trek, I walked past the Crazy Gecko, a cafe on stilts over the water's edge. A wooden structure decked out in different colours, vibrant greens, blues and reds. There were a number of hammocks and areas to lounge. It looked so inviting and relaxing. I made a note of it to return there at a later time. As I turned to continue on my way, someone called out to me.

'Hello stranger!' Beef and Eliza were lounging over the water's edge in hammocks. It was amazing to see them. A nice little reunion. We called the area outside of the Crazy Gecko Reunion Square in honour of that. I joined them and sat down to some food because I was absolutely famished. All they had given us on the bus was a condensed milk sandwich; a rather bizarre dinner option. They also gave out some Lao cookies. I would recommend the coconut ones, HGL was addicted to them and with good reason too.

I had my first helping of Shakshuka at the Crazy Gecko on Beef's recommendation. It was absolutely superb. What is not to love about a couple of fried eggs, a spicy tomato sauce and fresh baguette. Because of the French colonisation, baguettes are everywhere you go in Lao. The baguettes there, I was told were even better than those in Paris. Why? Because in Paris, industrialisation has meant all the baguettes are made by machines in big factories. In Lao, and especially on Don Det, they are made by hand. It was delicious and hit the spot. It became the meal of choice, certainly for breakfast, at the Crazy Gecko. You have to play it just right though making sure you have enough baguette to sauce ratio. You want a white plate at the end of it. Mop up all that sauce. It took me a couple of meals there to get it right. Beef was a pro at it, but then he had been there a few days already.

They gave me instructions on how to get to their guesthouse. I walked back up towards the 'town' if you could call it that. Ironically after all that, their guesthouse was the closest one to the pier. Literally, if I had taken a right off the pier when I arrived, there it was. However, had I not done the convoluted loop of the island I would have ended up staying at another guesthouse and been away from them. I even managed to get the bungalow next to theirs. The little bungalow structures were on stilts, hanging over the water's edge with hammocks on their own terraces. It was so tranquil, it was perfect.

I then rented a bicycle as that is the default mode of transport on Don Det. What is great about the island is that without any roads there is no traffic. This means that the air is clean and there is no noise pollution. All you can hear are the birds. I got a big clunky pink bike. There wasn't a huge amount of choice. I wish I had rented a mountain bike as I saw that at the next shop along. Oh well. Then again, that wouldn't be the same and not part of the romance of life on Don Det. Old clunkers are the way forward. The slow and suspension-free way forward. You want to feel every bump in the track and really connect with the island!

I regrouped with Beef and Eliza. We went for a cycle to the south and on to Don Khon the neighbouring island. There is a huge waterfall there, where the Mekong rushes through. It is ferocious and you can really see how powerful the river is. Cycling around on mud paths and getting soaked in mud, we were having so much fun.

Banter seemed to centre around the song 'The Scientist' by Coldplay. Eliza had never heard the song but told us of a friend she has known since childhood back in New Zealand. He had said that every time he hears that song he thinks of her. She didn't know the song so she didn't know what it meant. We told her it meant one thing and one thing only, that he was in 'luuuurve' with her. She didn't believe it but we were sure. Beef and I struggled through trying to sing the song to her to prove it, although, we didn't really know the words.

At first, Beef thought the lyrics were, 'Come up and meet me, tell me you love me, you don't know how lucky you are,' but I am not sure that is quite so romantic! In the end we sort of worked it out and muddled our way through.

'Come up to meet you, tell you I'm sorry,
You don't know how lovely you are,
I had to find you,
Tell you I need you,
Tell you I set you apart,
Tell me your secrets,

And ask me your questions,

Oh, let's go back to the start'

How could he not have been in love with her! That was quite a memory. The three of us on rickety bikes, bumping in and out of muddy puddles and getting soaked in the middle of the Mekong. Beef and I massacring a classic love song with our tone-deaf voices. Eliza going bright red, blushing, as she finally made the realisation her childhood friend was in love with her. Travel memories that stick with you. Ichigo ichie.

As we were cycling along the pathway, there was a big commotion with a lot of people gathered round. At first, we thought it was some kind of festival but as we got closer we realised that Mama of Mama's and Papa's restaurant had been taken ill. Beef and Eliza knew them because they had eaten there the previous night.

She was semi-conscious lying on the floor with a lot of people around her. No one knew what to do. This was the point Eliza, being a physiotherapist, stepped forward and turned on professional mode. It was such a contrast; one minute we were laughing and joking and the next minute she really took control of the situation. Beef and I didn't have a clue what to do. She deduced that Mama had had a stroke. She put her in an appropriate position, cleared some space around her and explained that she needed to go to a hospital as soon as possible. The closest one being in Pakse. Someone volunteered to take her and they set about moving her ever so gently and putting her on a boat.

It was quite a dramatic turn to the afternoon. It was so unexpected and such a contrast to the tranquil environment that we were in. It made me realise just how precious life is. It also made us realise just how far from civilisation we were. Ultimately we were on a silt island in the middle of the Mekong.

Once Eliza had done all she could, we continued our cycle onwards and came across the King Kong restaurant. It had a blackboard outside which proclaimed that it was highly rated in the Lonely Planet and served excellent homemade food. We decided to stop in for a drink.

This was pivotal for a couple of reasons. The first is that it was here that we met Adrian Robson. I had been telling Beef and Eliza about my meetings with Mr Sakura. I told them that I wanted to write a book about it.

Even at this point I still thought it was tongue in cheek. I didn't think I would ever actually get round to writing it. Would I go back to VV? Unlikely, not after that bus journey. Would I set up a Skype with him? Not sure given 1. how elusive and drunk he normally is and 2. the slow internet connection speed. Nonetheless, I was humouring myself and telling them I was going to write this book. I was telling them how I had met him, got on with him really well, celebrated his birthday and so on. From out of nowhere a curious Yorkshireman with no front teeth shouted out at us across the restaurant.

'Book? I've written a book.'

'Have you? Do you want to be my mentor?'

'No chance. I have written one book and one book only. Let me tell you all about it. It all started a year before the tsunami in Bukit Lawang in Indonesia. Banda Aceh was an area at the tip of Sumatra that was destroyed by the tsunami and caught the attention of the world. Bukit Lawang was the next town along and had been destroyed only the year before. There was a flash flood and the river there overflowed killing 600 people. It never made the news. I know this because I was there. I was one of only a handful of survivors.'

This was quite a traumatic experience for him, but he survived and it gave him the compulsion to write a book. He wrote about his experiences that day. Then kept on going from there, telling tales of his life on the road around Southeast Asia. He wrote it in ten days flat. He said he just locked himself in a room and wrote non-stop until it was done.

He was actually quite inspiring. He certainly inspired me. If he hadn't written his book, I don't think I would have written this one. And we just so happened to have the same name.

I asked him, 'Is this coincidence or fate?'

'I don't believe in fate.'

This was really the start of my journey to becoming an author. I was impressed and must have sat there transfixed hearing him talk about how he wrote a book. He said anyone can write a book you just need to write it. He showed it to me. It looked like someone's homework, printed out pages on A5. It was cut in a crooked manner and bound in a plastic cover adorned with clip art. I don't think he was best pleased with the final artwork.

'I wanted a Thai Buddha, I got a Chinese one. I wanted a magic mushroom, I got a toadstool. I wanted an orangutang I got a chimpanzee. That's an American passport with dollar bills, I asked for a British one and

Baht. She's not slutty enough to be a ladyboy in Bangkok. I wanted a Komodo dragon… I got a fucking gecko.'

What was fascinating about the story was that he had had conversations with publishers about getting his book published. They had turned him down, not because the book wasn't any good, but because they wanted to change the writing style to make it more legible. Adrian Robson refused to make even a single change to his work. They had wished him luck and ended the conversation.

His reasoning for not making changes, he told us, was because, 'I'm from Yorkshire and I am in yer face, the book is in yer face, that's why it's called 'In Yer Face' and I am not changing it. I write like I talk, like a Yorkshireman and so it's written like I speak, which you might have gathered by now is…'

'In yer face!' we all chimed together finishing off his sentence for him.

Ultimately, it could have been published but it never was. Instead, he sells it door to door and person to person; in railway stations, on the beach, in burger bars, in cafes and on the street. He says he even pretends he has a puncture on his bike and people will stop to help. Before they get a chance to ask about the tyre he will say, 'Do you like books?' and launch into his sales patter.

His sales patter was good and I bought a copy. Number 7,507. He said he keeps a list of everyone he sells them to and he signs each copy.

I tried to do his Myers-Briggs but he was having none of it. He didn't want to be labelled that was for sure. He had broken free from traditional western societal bounds and didn't want to be put into a box. I did get his profile in the end though. An ESTP and it was spot on. More on that later.

What a fascinating guy with interesting stories about Asia. Apparently, he is mentioned in one of the guidebooks too. If you see him, buy his book, it's a cracking read. He didn't want to be my mentor but he is the reason I wrote this book. Coincidence or fate? He doesn't believe in fate.

I think he only actually said that once, but it then became the catchphrase of the trip as so many coincidences (or perhaps fate) unfolded. It is only now that I realise not only is he the reason why I wrote this book but he is ultimately the originator of the title. Which in itself is another coincidence or fate. That is a nice nod to him. Cheers Adrian, I will tip my glass to you tonight.

The second reason why King Kong is of particular note is because this was the first time someone ever 'Wheeled'. After I bought Adrian's book, he left us alone and returned to the game of poker that was going on at the next table.

King Kong was a fabulous place. On the quiet side of the island with a great view overlooking the Mekong. It was a pirate's cove with a collection of random misfits from different walks of life. Ex-pats, people who had come on holiday and never left, even a miner who splits his time between Don Det and Western Australia. They smoke, drink beer, play cards and tell war stories of the scrapes they have gotten themselves into. The owner is a funny character. Big black curly hair. One of the nicest guys you will ever meet. He has done a great job of creating a fantastic place. The food is delicious too and he gets his coffee from a plantation on the Bolevan plateau on the Pakse loop. It smelt and tasted sensational.

The weather took a turn for the worse and a heavy storm came in. We decided to stay at King Kong and ride it out. We were going to be there a while so Eliza pulled out a pack of cards and proposed, 'What game shall we play?'

'Wheel or No Wheel?' declared Beef.

'Yes, let's play "Wheel or no wheel?"' I said, 'I don't know how to play it but I am willing to learn.'

They both looked at me strangely and then they looked at each other. Beef said, 'What I said was "one we all know" not "Wheel or No Wheel".'

I had misheard him. This was a fabulous, serendipitous coincidence or fate. I took the cards and dealt one to Beef.

'Wheel or No Wheel?'

His eyes narrowed. I just nodded at him encouragingly. Eliza didn't have a clue what was going on. He picked up the card, studied it for a moment and then reluctantly decided, 'No wheel,' and discarded the card. I then dealt him another one which he looked at and seemed much more pleased with. 'Wheel,' he stated with a level of authority belying his uncertainty. This took me by surprise as I was unsure what would be a good card to 'Wheel' on myself.

I then turned my attention to Eliza, dealt her a card and asked, 'Would you like to wheel or not to wheel?'

'No wheel.'

Another card.

'How many chances do I get?'

'Three.'

'No wheel.'

I dealt her a third and final card face down in front of her. She went to look at it but I stopped her.

'Your third card,' I informed her, 'according to the rules of the game has to be played blind. Neither you nor anyone else can see it until the big reveal.' She recoiled as if she knew this already and was kicking herself for being such an amateur.

I then dealt myself a card. It was an Eight of Clubs. I decided not to wheel. I then got a Nine of Diamonds and again decided not to wheel. I dealt myself a blind.

Beef unveiled his card to be a Queen of Clubs. Eliza a Jack of Diamonds. I had a Two of Clubs. Beef fist pumped the air. 'Winner.'

'Hold on,' I said, 'in "Wheel or No Wheel?", low card wins.'

I take my two and place it to the side next to me face down and take their cards and add them to the other discarded non-wheeled cards from earlier. 'That was one round. We keep playing until we have gone through the pack. The winner is the person who has the highest total when you sum all the cards you have won with. Low card wins a round, Aces are low, but highest sum wins overall.'

And that is essentially how the game was born. We added a couple of rules as we went. If two people had the same card, they would cancel each other out and the next lowest card would win. The Ace of Spades is the all conquering. It cannot be beaten. It cannot be cancelled out with another Ace. When you win with an Ace of Spades you get to steal someone else's face down winner card. The Seven of Hearts was another magic card. This meant that if you won with it, you could double another of your face down cards at the final reckoning. Hard to win with a Seven of Hearts but if you take the risk and win, you get rewarded for it.

'You always wheel on a Seven of Hearts.'

Another final rule to cap it off is that you rotate dealer each round. One time in every game the dealer has an added opportunity if they are playing blind i.e. they haven't wheeled on their first two cards. They can steal a card from another player. They then swap their blind with the other person. This can work for or against you as you may have been given a good card and equally you may steal a bad card. It is important to keep track and see what reactions people have when they get their cards. Note whether or not they are worth stealing.

Those are all the rules. It is a simple game but has some incredible twists and turns. For instance, when Beef and Eliza both got a Three each, cancelling each other out. I had a blind which turned out to be a King and I won. That was fortunate and unlikely but it netted me 10 points in the final reckoning (Face cards count as ten).

Getting the Ace of Spades first round is great because you automatically win, however, there is no one to steal from. One time the three of us all cancelled each other out with Aces. In that instance paper, scissors, rock determines the winner.

There are also a number of strategies that emerge too. Beef realised instead of winning multiple hands with Aces and Twos (which he termed breadwinners) he only needed to win once with a Jack or Queen to be in with a good chance of winning. He only wheeled on high cards.

I would recommend that you play it. It is a fantastic game. We really enjoyed it and got quite animated playing it. The rules are simple. Anyone can play. Easy to learn, difficult to master.

The full rules can be found in the Bonus Features at the back of this book.

From there we rode back in the dark which was treacherous. Negotiating dark, uneven muddy paths with massive puddles was not easy. On an old clunky bike, it was even worse. It was bumpy, it was wet, it was difficult. I used my phone for its torch. Cycling one handed made it even more precarious. There were a couple of falls in the mud but there was no shame in that.

Adrian Robson told us about a great little curry place in town to eat at and that he goes there most nights. We opted to try it for dinner. I was exhausted because I had been up for over twenty-four hours. It was unbelievably only that morning that I had arrived on Don Det. I had a terrible headache and was in need of sleep. It took ages for the food to come. As soon as it did, I ate and had to excuse myself and go to bed. I apologised. The long bus journey and all the excitement of the day finished me off. I was in bed by 8.30. As soon as my head hit the pillow I was out like a light.

CHAPTER EIGHT

OPPOSITES ATTRACT

I woke up fully refreshed. It was a lovely morning and I sat out on the balcony next door to Eliza and Beef's. I sat on my hammock, Beef on theirs and Eliza in the space in between. The sun was shining, the river was flowing. Having a chat. Really chilled. Eliza had brought with her a selection of special teas so we drank those.

We played a game of, 'How many of the 24,000 islands can you see right now?' Every time we counted we got a different number. I had 63 the first time and 65 the second. It also raises the question of what exactly constitutes an island? Is a big bush coming out of the water an island? What about larger pieces that are attached to a bigger island but there is definitely water in between.

Beef questioned as to why I called it the 24,000 islands. Apparently, it is supposed to be called the 4,000 islands. I felt a little disappointment. That was only a sixth of my expectation. When I looked it up, there isn't actually even 4,000. That was just an arbitrary number that was sufficiently large because there are lots of islands. Therefore, there might actually be 24,000. There are certainly at least 65.

We got on our bikes and cycled over to the Crazy Gecko for breakfast. En route we saw Adrian Robson by the side of the road with what appeared to be a puncture. There were two travellers helping him out. I looked over to Beef knowingly.

He winked, 'I wonder if they like books.' The three of us all laughed.

A few minutes later we were on our favourite hammocks at the Crazy Gecko. Geordie Sam and HGL came past - and we reunited with them - where else but Reunion Square as it turned out!

'I've just bought this book,' Geordie Sam exclaimed.

We realised they had been the ones chatting to Adrian Robson.

'Did you manage to fix his puncture?' Eliza asked cheekily.

'Actually, we never did in the end. That's the reason we stopped to help. Should we go back?'

'I think he'll be fine,' Beef laughed. He explained that we had met him the day before and that I too had bought a copy. We had a good laugh about it and they joined us for breakfast. We got to know them really well that day and I felt a strong connection to the Geordies. I liked them anyway from my first encounter, but now we were book buddies too!

They were your archetypical post-university backpackers on a gap year. However, there was something a little different about them. They were young, bright, positive and optimistic. They had a can-do forward-looking attitude as if anything was possible. They were enjoying every single moment. They were so present and in the moment. I thought this was just youth but it was more than that.

They told us about a mind-boggling travel story which I think pretty much explains their positive energy. They had graduated from University and had saved up their money to go on a dream backpacking trip away for a year. They were going to go to Asia, Australia and South America. Everything was all set. They had their flights booked and they were ready to go.

That was until the night of their graduation ball. HGL a keen gymnast, after a few drinks, decided to perform some of her acrobatic moves. Not the best idea given all the celebrations and free flowing booze. She attempted a convoluted jump and it was too much. She landed funny on her arm and broke it. Fortunately, she was so drunk she said she never felt the pain but the problem was that this was three weeks before their due departure date. It meant they had to postpone their trip and maybe would even have to cancel it altogether.

'Well in some ways, it's a "coincidence or fate", right? Because if you hadn't broken your arm you would have gone through Asia already and you would never have met us,' I offered.

'It's even more "coincidence or fate" than that!' Geordie Sam recounted. The day he went into work looking at his calendar and realising it was the day they were supposed to be flying out was a bad day. He was in a rotten mood. He was irritable and every little thing was making him more and more annoyed. HGL felt even worse. Equally disappointed not to be flying away on an experience of a lifetime, she had the added guilt, knowing it was her drunken fault that she'd let Geordie Sam down too.

Geordie Sam was sitting at his desk having had a particularly unproductive morning and not wanting to talk to anyone. He flicked open a news website thinking maybe that would take his mind off things. He stops dead in his tracks. Like a statue. Instantly a film of cold sweat coats the back of his neck. He gets goosebumps all over his entire body and a rush of warm adrenaline shoots up his spine. 'Passenger jet shot down over Eastern Europe.' He looked at the flight number and he can't quite believe it.

'That was our…'

His phone rings and he realises he hasn't taken a breath for some time. It is HGL. 'Have you seen the news?' She was in shock too. They both were. As were me, Beef and Eliza when they told us the story. Listening to him we got the goosebumps and the adrenaline as well. We were sat on the banks of the Mekong, mouths agape. We couldn't quite believe it either. If HGL hadn't broken her arm, in that drunken foolish way, then they wouldn't have been sat in the Crazy Gecko with us eating Shakshuka. They would have been on that flight.

After the seriousness of our breakfast chat, we chilled out on the triangle cushions and hammocks overlooking the river. We taught them the rules of 'Wheel or No Wheel?' and they loved it. The drama is even greater with more players as more cards get cancelled out. I remember saying to Geordie Sam, 'Today is going to be really a good day.' He nodded in agreement, 'It's going to be unreal.'

I got everyone to do their Myers-Briggs and tested the Geordies. I discovered Myers-Briggs several years ago and it changed my perspective on people and life in general. It is a personality test that we did as a team in my old company. The idea being to identify what people are like and therefore how they behave in work situations. By mixing up skill sets and getting a good balance you can play to people's strengths. It works outside of work as well and just a great way of understanding how people behave. You can do courses and questionnaires but I just boil it down to four questions that can pretty much give you you a steer of your Myers-Briggs profile in only a few minutes with a high degree of accuracy.

These questions are as follows:

1. Do you prefer to sit in the corner and think it out (I) or do you need people to bounce ideas off (E)?

I - Introvert or E - Extrovert

2. Do you like the big picture (N) or the detail (S)?

N - Intuitive or S - Sensing

3. Do you consider people's feelings and aspire for harmony (F) or does the task have to be done on time with a focus on results (T)?

F - Feeling or T - Thinking

4. Do you write lists and stick to them with meticulous rigour, ticking off as you go (J) or are you more fly by the seat of your pants, see where the day takes you (P)?

J - Judging or P - Perceiving

Add your four letters together and if you are like me, for example, you'd be an ENTP.

Extraverted. Intuitive. Thinking. Perceiving. "The Visionary".

Answer the questions and check the Bonus Features at the back of this book to find out what you are.

When I read the profile of an ENTP I thought this is actually exactly how I am and explains so much. In short, an ENTP is a visionary, full of ideas who likes to bounce them off other people. I am very much into people and debating. Love to converse and tell stories. Fly by the seat of my pants. Am an ideas person, coming up with lots of them and enjoying the analytical nature of solving problems. The problem is I am not good at acting on them. Whilst I am very good at coming up with the idea the follow-through takes more effort. This book was a classic example! It was a great idea and it took a while to finally act on it.

With Geordie Sam and HGL, we asked them to do it for each other. So one answered on behalf of the other. This did two things. Not only did it get us to work out each of their personality types, it also got us to see how well they knew each other. The results were astonishing for two reasons. Not only did they manage to identify each other's personality type spot on, they were exact opposites. Perfectly balanced, they cover all of the bases. No wonder they are such a perfect couple. Opposites attract.

Geordie Sam was an INFP, The Idealist and Honorary Geordie Liz was an ESTJ, The Guardian.

He was introverted, likely to sit on his own and reflect. A man of few words but when he did speak it was usually wise and well thought out. This is because of his intuitive and perceiving nature. HGL, on the other hand, was more outgoing, likes people and is a great conversationalist. She likes order and detail.

This was a truly magical moment. Everyone was surprised by the coincidence and now nicely it fit. Or was it fate?

'I don't believe in fate,' Geordie Sam exclaimed mimicking Adrian Robson.

That evening we went to Sunset Bar to watch the sunset over the islands. The sky turning all manner of pinks, purples, oranges and red, we noticed all the crazy scrawlings people had written on the walls. Such nuggets of advice included:

'Don't drink and drive, don't smoke and fly, TAKE LSD AND TELEPORT.'

'The sun burnt every day, it burnt time. Cas.'

'Your'I [sic] never gonna find yo-self but you might touch the ether.' They obviously had.

'Only dead fish follow the stream - RR OhighO'

'If a mad dog bites you… don't bite back for then you will be mad too.'

'Happiness is real, only when shared. Mao'

'The world is small but the family is BiiiiiG!!! See you guys soon! One Love'

It was this evening that things got a little awkward between Beef and Eliza. I am not sure why. They were having some heated words at the bottom of the stairs to Sunset Bar. I carried on with the Geordies so as not to intrude. Eliza eventually caught up with us and Beef was nowhere to be seen.

Eliza and I went to Reggae Bar and hung out there with a few of Don Det's local ex-pats whilst Beef finally rejoined proceedings and played Geordie Sam at some pool. The next day Beef and Eliza left. That was, in fact, the last time I ever saw them. Beef was heading to Siem Reap, Eliza to Phnom Penh.

There is a Japanese Zen phrase, ichigo ichie, which translates as, 'Treasure every encounter, for it will never recur.' As we were saying goodbye I remembered it. Our time on Don Det was like that. It never happened previously and it won't happen again. Those people, in that spot, at that time, in those set of circumstances. No one ever knows what the future holds and so every moment, every meeting is unique and precious, and won't happen again. It is a good phrase to live by and teaches you to enjoy every moment and appreciate it for what it is. For it will never repeat.

After the emotional goodbye with Beef and Eliza, I went to Mama's and Papa's for lunch. It was good to hear that although not back on the island, Mama was, in fact, ok and recovering in hospital. In her absence, her sister was manning the kitchen. The English guy holding the fort at the front of house said that she didn't understand what he said, so it was pot luck what food you got.

I ended up with a rather delicious falafel and sweet potato wrap. I then went to Crazy Gecko bar with the Geordies and bumped into Sally, Claire and Baz, also, as it happens at Reunion Square. Seeing travel buddies whilst you are touring around is like meeting old friends. We chatted nonsense and had good banter. It was a Sunday afternoon and it felt that way. Every day felt like a Sunday afternoon on Don Det. We played 'Wheel or No Wheel?' and they, of course, absolutely loved it. We did everyone's Myers-Briggs.

The next morning the Geordies and I left the island. Waiting for the boat we bumped into Jesse J who we had met on the bus ride down to Don Det. We had arrived on the same boat as he had. We left on the same boat. We had all intended on staying two days. We had all been there a week. Don Det does that to you.

Overall Don Det was great. It had really opened up my creativity. I had invented a card game and I was inspired to write a book. I had laughed so much and connected with so many great people. I really felt myself and in touch with nature. Above all, I was in awe of the river. The full force of nature at it's most beautiful. Always present, always flowing.

It just feels so good thinking about it. How simple life can be. How easy life can be. Living on a silt island in the middle of the Mekong for a week. It was such a carefree time. Looking at the faces on the pier you could tell everyone felt the same. There was quite a sense of community amongst everyone. Everyone had a knowing sense of bliss.

CHAPTER NINE

THE PAKSE LOOP

We all left Don Det with a strange feeling. It was sad to be leaving but the time felt right. Sally and Claire were going to Burma, Jesse J to Vietnam and Baz was heading to Cambodia.

The plan for the Geordies and I was to go back to Pakse. We were going to get dirt bikes and do the Pakse loop as per the Lexpedition's original instructions. This was, in fact, the reason he loved Lao so much in the first place. It was certainly strange being back in a city environment. The concept of roads and cars really took me by surprise after a week of mud tracks and bicycles. Everyone missed Don Det instantly.

We found a guesthouse and Geordie Sam and I went in search of dirt bikes. HGL planned our route and researched the Bolevan Plateau. Geordie Sam had never done a bike road trip before and neither had I. It only just dawned on me at that moment that it was a road trip and I love road trips. Geordie Sam was a bike expert, at age 14 he was a UK National Dirt Bike Champion so I was in good company.

We got our bikes and hit the road. There is nothing quite like it. As soon as we got out of Pakse city limits the scenery of the Bolevan plateau takes centre stage and it is very special. This is not an area that is highly visited as it is so far away from anything or anywhere. There is no big airport close by and so there are not truckloads of tourists coming through. The only way to do it is on a bike. We saw very few tourists for four days. The main ones being in Tad Lo and that was right at the end.

The plateau is lush greenery in all directions. Rich in flora and fauna. There are a number of beautiful all-natural waterfalls on the route. These require some hiking to get to and we went to several. Our days consisted of road tripping on the bikes, finding a hiking route and visiting waterfalls. Sometimes we would bathe in the naturally formed pools to cool down from the tropical heat. We enjoyed the vistas and being amongst nature.

We would then find a tiny village to stay for the night. We'd find a local bamboo hut restaurant and sign language what we wanted to eat. They didn't understand us and we didn't understand them but we had fun and laughed with the locals. Our favourite moment of the whole road trip was in 'Ikea Valley'. We took a selfie of the three of us and the most beautiful view of an entirely untouched valley - a garden of Eden. It was quite something.

I said to the Geordies, 'Take it all in now, for in 20 years this will probably all be Ikea furniture.' I sincerely hope this premonition doesn't come true. The view was spectacular and Lao is one place in the world where it seems nature is winning.

I had decided I was definitely going back to Vang Vieng now. I was going to interview Mr Sakura and get what I needed information-wise for the book. I was going to do a DJ set at Sakura too. It had been my dream to be a Superstar DJ since I was a child. Don Det had inspired my creativity in all aspects and music had been yet another one of them. Geordie Sam and HGL helped me. They downloaded some software for me and brought me up to date with where the scene had got to.

Whilst the scene had evolved, one thing for me stayed the same. The secret to being a great DJ lies in the way you build up your set, taking the audience on a journey. Creating waves of emotion, and moments to remember. Whilst everyone and their dog seemed to be a DJ these days, and some of the songs sounded like your gran had left the hoover on, this notion seemed to be missed by a lot of the new school of thought.

We started crafting a set of old school timeless classics, modern house with funk and a few mainstream party tunes to keep a broad appeal. Something for everyone and something guaranteed to have the crowd loving it and wanting more. This was to complement the already amazing Sakura experience by adding to it with an emotive soundtrack. When I used to play out in Manchester I became famous for my encores. One, two, three… as many as the crowd wanted to keep them satiated but also leave them just wanting that little bit more.

It was another creative four days as we downloaded all these songs at night and then listened to them on the bikes in the day time. This was such a liberating experience. The wind in your hair with no other traffic on the road. The glorious scenery on either side. Incredible green hills, completely devoid of human activity. Stopping off on the side of the road at regular intervals we would go for a hike and get a selfie with a natural waterfall.

Unfortunately, we could only stay for four nights which was a shame but such is life on the road. When you are a budding author and superstar DJ in the making, you have things to do. The next day we were back on the road for the final leg back to Pakse. I was excited because I knew I had to be back there for 7 p.m. in time to make my bus back to Vang Vieng. I had a feeling of trepidation in my stomach and was excited about the thought that by the following day I would be back in Double V.

When I had told the Geordies that I was going back to VV you could see their eyes narrow with envy. They were excited for me but they also didn't want to miss out. Especially if I was going to be DJing as well. They had been particularly instrumental in helping me build the set. Their plan was to stay in Pakse that night and take the first bus to Siem Reap the following morning. I felt a little sadness that we would be saying goodbye after this incredible adventure we had been on. It had been another perfect ichigo ichie.

The Pakse loop had been one of my best experiences ever and was so happy to have shared it with them. I would never have been able to do it without them. Geordie Sam for teaching me to ride a dirt bike and HGL for creating an itinerary that took in all the best sights.

As we were on the open road back to Pakse they pulled over in front of me. I thought this was strange because there wasn't any point of interest around. They waved for me to stop. I looked at them confused. HGL took her helmet off and said, 'We're coming with you to Vang Vieng.' I was surprised, ecstatic and worried all at the same time. Really surprised because I knew they had friends to meet in Cambodia. Ecstatic because I knew I couldn't do this without them. Worried because what if we went back and Mr Sakura had forgotten everything. Forgotten about the book, the DJ set, everything. He might even have closed for the rainy season. The last I had seen of Sakura, the day I was leaving VV there was a pile of rubble outside as if they were doing a refurbishment.

I had spent the last two weeks in the moment that I hadn't stopped to think. I was suddenly concerned. I took a moment and relaxed. If this trip had taught me anything it was that all I had to do was put my faith in the moment. Follow my instincts and my energy. I knew it was all going to be fine. At that point, I just felt a complete sense of calm and acceptance. What happens happens.

We got back to Pakse and everything ran smoothly. We left our bikes at Sout Chai travel company and got on our minibus to the bus station. We had top bunks on the bus and played some 'Wheel or No Wheel?' before calling it a night. No one slept particularly well. There was a ferocious electric storm outside which kept us up. It was spectacular to look at mind and we were so excited about what the next few days would hold.

There were not that many westerners on the bus but plenty of Koreans. We arrived into Double V just after midday and it was pouring with rain. The streets were empty. We were the only ones who got off the bus. The rest stayed on to Luang Prabang. I looked around with dismay. The town was dead.

CHAPTER TEN

RARE FIND

We checked into the guesthouse opposite The Corner Cafe where I had had my first beer with Armstrong, Eliza, Beef and the Welsh girls on our first night in VV. We got a good deal and I set about looking for Mr S.

I went to the bar and found one of the barmaids. I asked her where Mr Sakura was and she said that she hadn't seen him all day.

'He is usually up by 3 p.m., but I don't know, last night he had a lot of tequila.' That statement pretty much sums him up.

I went and knocked on his door. I rapped on it several times and was about to give up when I heard his muffled voice, 'Yeah?'

'It's Robson.'

'Oh, uh, can I get five minutes?'

'Do you want an hour?'

'Yes!'

I was worried it was going to be a little awkward. Remembering the last conversation we had had was on his birthday and he had been pretty drunk. I wasn't really that sure if he was into the book idea for real or if he would even let me DJ. He had a business to run after all. He was fairly busy that afternoon but he seemed cool about the DJing and the book although I really couldn't tell. He didn't seem that committal.

The Geordies and I then hit it hard in Sakura. Why not? It is after all the most famous bar in Asia. We had our free drinks during the happy hour and got some T-shirts. They had some new colours I hadn't seen before - grey, black and purple. But still no blue.

Geordie Sam and I were standing at the back bar and when we purchased our vodka drinks were asked which top we wanted. We looked behind the counter and on the shelf at the back, underneath a pile of T-shirts, I saw a white one. It had the pink trim on it. The highly rare limited edition T-shirt that had been made for Canada day. I asked them to pull it out and there were two more with it. Unbelievable. We quickly bought another couple of vodka drinks and secured the three. There were only three left. They had just been discarded behind the bar. Geordie Sam, HGL and I were the only ones to have them.

We inhaled plenty of balloons. Of course, we danced on the pool table. A trip to Sakura is incomplete without it. I developed a new batch of Korean fan club members who Geordie Sam taught my name. They took great delight in chanting as I gyrated on the pool table.

'Rob-son, Rob-son, Rob-son!'

At midnight when it was time to go, Mr Sakura came and found me and said, 'Tomorrow, 1 p.m., come here and we will do the book.' I was so happy at the realisation it was actually going to happen. I had to get him to say it again just to be sure I heard him right. I got myself a Laughing Cow baguette to celebrate. I was still so hungry I went back for another one. Wheel!

REMINDER FROM ROBSON IN ROME: I hope you're enjoying the story so far and looking forward to Part Two. By now I think you probably have a picture in your mind of just how beautiful Lao is. If you haven't already, consider posting a photo on social media of you reading this in an equally random or beautiful location tagging @RobsonDob and using #SakuraBook. Then you can be added to the map and also check out where others have been reading it too! Best photos may even win a prize of a limited edition T-shirt.

Share a photo and be part of the story @RobsonDob #SakuraBook

PART TWO

IN CONVERSATION WITH MR SAKURA

CHAPTER ELEVEN

MR SAKURA ON SAKURA BAR

ROBSON: I'm here with Mr Sakura, it's quarter past one. What day is it? I don't even know what day it is.

MR SAKURA: Monday. Monday 27th of July.

ROBSON: Wow. How remarkably coherent of you.

MR SAKURA: I gotta stay on it a little bit.

ROBSON: You are a business owner after all. Do you see yourself as an entrepreneur, as a business owner?

MR SAKURA: Ah, I don't know. Not really. I guess at some point, but right now, I am just going with it. Things have been working out this time round. I've taken a few stabs at it with Sundowners Bar first, then Happy Bar and now Sakura. It has been an evolution of those bars which came along before it.

The vibe was very different at Sundowners Bar; it was more of a chill-out place. It was where you could sit in a hammock, do mushrooms and not have to worry about hanging out with all the drunk people. Sundowners was the drug bar. Happy was more about the party. When we moved to

Happy we were competing with the nightclubs in town. We were playing more dance music and we had a much bigger area to work with.

ROBSON: That was the evolution you were talking about?

MR SAKURA: Yes. Sakura is the next step, it's a step up from Happy. This is what we were working towards and it's in a much better location.

ROBSON: Is that all it is?

MR SAKURA: I think so. It's the same people and the same kind of vibe.

ROBSON: Did you do the drinks promos? The T-shirts? Laughing gas?

MR SAKURA: T-shirts, everything; it has all just moved to Sakura. If we could go back in time two or three years and I took you to Happy Bar it would feel the same as Sakura, only in a more basic format. It was on the island in the middle of the river so it was more open and we were affected by the weather more. On rainy nights we wouldn't sell anything so this part of the year is new to me. Having a business in the wet part of the year is awesome.

Before we would try to make as much money as we could before it started raining. One year we tried moving to a couple of different places. We always tried to find a venue that would take us and as soon as we did, they would find a reason to kick us out. We always screwed up at the end of the year by getting shafted on the rent or something like that.

This time we are doing everything properly. We have rented it and paid up for five years. It is going to be here for another five years for sure. That is why I am putting so much effort into it and trying to build it. Five years of smashing it.

ROBSON: I always had a taste for a good party and have been around the world partying; Ibiza, Vegas, Miami, Hollywood, London, Marbella, Bangkok, Hong Kong, Sydney, Manchester, wherever, all over the world... Koh Pha Ngan, and I have never, ever, partied like I did here. What is it about this place that makes it so special?

MR SAKURA: I really don't know. I think being welcoming… it's one of those weird ones, I don't know why people are so into it here… the strange thing is there's never a fight. Sometimes it will get a little rowdy… well every night they get a little rowdy, but the hostility level is almost at zero which is the very opposite of where I am from. In Canada, there's a fight every night and you get popped from all over. I love it here when it's packed with Koreans and Thai and Lao and every western country you can think of… and Brazilians, South Africans… fuck and even Swedes and everyone is just real chill.

ROBSON: I counted the nationalities and I met 30 different ones in my first 48 hours

MR SAKURA: That's amazing man, that's cool, I like that.

ROBSON: You want to see the list?

MR SAKURA: Yeah.

ROBSON: I think I have a couple to add because I met an Austrian and a Belgian yesterday.

MR SAKURA: Oh nice.

ROBSON: My favourite one was Kuwait.

MR SAKURA: Oh what? Wow!

ROBSON: Then Brazil. A group of 15 women; that was amazing. That was in the first tubing bar I went to. I bumped into these 15 Brazilian women and I thought, 'If this is tubing, then tubing is pretty fucking good!'

MR SAKURA: Yeah. Haha. Nice.

ROBSON: Peru, Italy, Israel, Iraq, Brazil, Kuwait, Turkey, Korea, Norway, Germany, France, America, New Zealand, Holland via Canada, Argentina, Spain, Wales, Denmark, Poland, Scotland, Australia, Real Canada, Japan,

Lao, Sweden, French Canadian… French Canadian they called themselves their own nationality.

MR SAKURA: Yeah, they do that.

ROBSON: He said I had to call that out as a separate one. I said, 'Aren't you just Canadian?'

He said, 'No, I'm French Canadian. I am not French and I am not Canadian.'

I asked him, 'What passport do you have?'

MR SAKURA: 'Canadian.'

ROBSON: Thailand, Chile, Switzerland, South Africa and I will add Austria and Belgium.

MR SAKURA: That's fucking incredible, that's huge!

ROBSON: So you have all of these nationalities in one place and then you put on the best party in the world. How did you do that? Did you create that? Is there a formula for that?

MR SAKURA: Giving out free drinks tends to get people in a pretty friendly mood you know… and I think it is just Lao… you aren't allowed to really fight or even yell in Lao and it really rubs off on people. By the time they get here, after all their shitty buses, they've had to deal with the Southeast Asian pace of life. I think you get a lot more patience. That and you meet a lot of people on the way who you have had a shared experience with. You see the same people and maybe you haven't talked to them yet.

It's easy to start up a conversation; 'Hey weren't you in Chiang Mai a few weeks ago?… Yeah, yeah, ok, cool, fuck, have you been to Siem Reap, have you seen Angkor Wat?…'

'No, no I'm heading that way…'

'There's this guesthouse you must stay at.'

It is also a nice part of the loop where it is not super busy, not like Bangkok or Koh Pha Ngan. Those places are quite anonymous and when you are there you tend to hang out with whoever you came with. To get here, you probably took the slow boat from Chiang Mai to Luang Prabang. They would have spent two days stuck on a boat together. You get lots of shared experience that way. By the time people get here, they have learned how to figure out how to deal with people and make new friends.

If you've come this far you kind of have to put the effort in and if you put the effort in you appreciate it more and you have greater gratitude. Everyone who made it this far had to go through some sort of trying situation.

ROBSON: A six-hour bus ride without air conditioning for a start. That makes sense and they're a long way from home and they're away from their friends. These are their new friends.

MR SAKURA: And you don't have your friends judging who you're talking to or any of that kind of cliquey stuff that happens at home.

The other thing is the Koreans. The Koreans know how to have a good time. I was in Korea for four months a few years back and they like to fucking party. I think it's probably the first time a lot of westerners have seen Koreans party.

ROBSON: I guess Koreans are your number one customer?

MR SAKURA: Right now, definitely… but in high season we get lots of Aussies, lots of Brits, lots of Swedes, lots of Canadians… always lots of Canadians. Americans not so much, but if they come they are usually West Coast. West Coast is real chilled. Real chillers.

ROBSON: One thing I noticed is that because the ceiling is so high it makes the bar really open.

MR SAKURA: That was one of the first things I did. It means you have to talk to people. A lot of bars are like that where I am from. I always pay attention to shit like that when I'm at bars. A lot of the bars back home have the dance floor in the middle and everyone walks around it in a circle in one way or another and you've always got to talk to people. You always bump into people and there are a few bars around the outside so everyone has to face each other. You're always all looking in, into the dance floor, into the centre of the bar. I would also like higher tables around the outside so the people sitting down make eye contact with those walking past. If you sit low down and someone walks past you, you don't have to acknowledge them at all. If you're sitting higher and someone walks past, you catch their eye. That forces interaction. I also only like benches around the outside of the seating area and moveable chairs on the inside. It is set up in a way so that you have to be close to the group next to you.

There are lots of bars in Vancouver with really big long beer garden style tables. If you want to sit, you end up sitting next to someone you don't know and you end up talking to them. That's what I am trying to do with the seating which forces people to hang out, integrate, talk together, get drunk together and have something in common. A shared moment.

ROBSON: Ichigo Ichie?

MR SAKURA: Yes! Exactly!

ROBSON: So you are orchestrating the vibe? You are thinking how can I make this the greatest, most interactive bar?

MR SAKURA: Yes, I want to make it better and more fun all the time. Just like with the beer pong. We were the first people to bring beer pong to Vang Vieng when we were at Sundowners Bar and that just took off. If I am proud of one thing, I'm proud of being the first one to bring beer pong here.

ROBSON: That's always a great game; you can flirt, you can chat and it's competitive. It is a drinking game too so it gets you a bit tipsy.

MR SAKURA: And you have multiple players.

ROBSON: Anyone can play; it's just lobbing a ping-pong ball up. Did you struggle to find ping-pong balls here?

MR SAKURA: At first I did. Then, once I started buying a lot of them, the shops around here started stocking them You will see bags of ping-pong balls… but no paddles.

ROBSON: You basically created the ping-pong ball industry in Vang Vieng.

MR SAKURA: I don't think there was much ping-pong being played here. The beer pong table used to be part of a real ping-pong table. The other parts are long gone.

ROBSON: They are busy with people playing on them all night long. It doesn't get moved out of the way?

MR SAKURA: Sometimes it naturally does but I just try to keep an open path. Mostly just for me because I am always going back and forth between the bar at the back and the one at the front.

ROBSON: I noticed that yesterday; one minute you're here and then you're there and now you're back over here again.

MR SAKURA: I am just making sure everything is good and everyone is set up right so the bartenders don't have to leave their station. I like to be in control because then I know everything is being taken care of. I have a real hard time distributing that bit of control. I like to have everything run how I want it to be run, my way, because I have good ideas.

I am always picking up on a few things wherever I go. I look around and think 'That's a good idea, I could use that.' I've managed bars and restaurants back home and a pizzeria that does amazing pizza… Two Dudes Pizza.

I learnt a lot about how to run a big business with good staff relations when I was there. Our boss was just amazing to us and he really supported us. That made everyone work harder. We really built the brand and got people on board as well; that pizza place was really great. I really learned a lot. Two Dudes and a Pizza Place back home where I'm from. They won best pizza in Canada twice.

I used to say 'Hey Casey, I'm back for 6 months can I get my job back?' He would always take me back and he used to ask how I was getting on with business here. He wanted to know what I was doing with advertising and he was great to just bounce business ideas off. He taught me a lot about advertising and managing. He's kind of my mentor.

He's Canadian. Casey Maxx. He started Two Dudes and a Pizza Place from nothing. It is a real popular place where all the kids go to get pizza. They deliver to all the bars and do a lot of cross-promos with bars during sporting events. He's everywhere. It is the same thing I want to do with Sakura; be everywhere, so you see it and you're almost forced into becoming a part of it. You just can't avoid it.

CHAPTER TWELVE

MR SAKURA ON MR SAKURA

ROBSON: Would you say you are naturally a party animal?

MR SAKURA: I think so, yeah, definitely.

ROBSON: As a kid?

MR SAKURA: Yeah, growing up, at high school, at university…

ROBSON: Would you say you were more of a mentalist than average?

MR SAKURA: Yeah, I think my friends and I liked to party a lot. I grew up in a university town and every night of the week there was a different special on in each bar. I got my first fake I.D. when I was 16. The drinking age in Canada is 18. I started drinking when I was 13 or 14, and it just continued from there. When I was in university for the first couple of years it was a write-off. Then when I came out here, to Southeast Asia, I partied a lot and then I moved back to a party town called Banff. I think it's always just been that way.

ROBSON: Do you call it a job?
MR SAKURA: Yeah.

ROBSON: It seems a fun job.

MR SAKURA: It's like my hobby, my full-time hobby.

ROBSON: This is an interesting thing because I think to the outside observer, you're this party animal, you run this bar, you rock up, you're hammered...

MR SAKURA: They think I've got the easiest job in the world!

ROBSON: And that it all just runs like clockwork. One minute you're stoned in a ditch, and then you wake up the next morning and you're like, 'What the fuck happened last night?'

Then you're back on the mic. 'Hey everybody! Drink triple, see double, act single...' That's the image that is portrayed of you and that's probably what people think of Mr Sakura. I guess that is probably what you want as the owner of a bar.

MR SAKURA: Oh yeah, for sure, that is what you should want, people thinking, 'That guy he's always having fun.'

ROBSON: Doesn't have a care in the world?

MR SAKURA: Yeah, that's what I want to be...

ROBSON: That is half the thing with the book; everyone at some point in their lives thinks, 'I want to own a bar in Southeast Asia. I could do that. Be Mr Sakura. Get drunk and sleep with loads of women each night.'

MR SAKURA: That's the dream life.

ROBSON: That is why people will want to read about it?

MR SAKURA: Exactly. They don't want to hear about... I do absolutely everything!

ROBSON: They would also think you were some dropout from Canada, couldn't find his way in his own society and that is why he's here. For me, that's why this is so interesting.

MR SAKURA: It is a bit of a different story isn't it?

ROBSON: Yeah, it is.

MR SAKURA: I just figured out a couple of things.

ROBSON: That is why I think the story is so interesting because it is not like that at all. And I got a sense of that talking to you last week, when you were saying that actually you're almost an ambassador for Vang Vieng and you want to help put the place on the map.

MR SAKURA: That's my 'Welcome to Vang Vieng'.

ROBSON: Exactly.

MR SAKURA: I try to say that to everyone, anytime I haven't seen them around, 'What's up? You just got here? Welcome, what do you think of it? Have you checked this out yet? Have you done that? Yeah, stay longer, it's a wicked place. Oh, you haven't done this yet, you have to,' and that is getting people to stay longer.

CHAPTER THIRTEEN

MR SAKURA ON TUBING

ROBSON: Tell me about tubing? You didn't start tubing?

MR SAKURA: No, no, it wasn't me.

ROBSON: The guy who invented it had a farm. Is that right?

MR SAKURA: Yeah, yeah, his story is that he's got the organic farm right up where they start the tubing. It is next to Life Bar on the river. People would stay there and work on his farm for free and he would give them accommodation. He'd lend them or rent them tubes to come into town or if they just wanted to float along the river as an afternoon activity for them to do.

ROBSON: Basically tubing is a Laotian sport because they've got no money but they've got loads of tractor wheels? I love that it is so eco in a way because it is recycling something that would otherwise be disused?

MR SAKURA: Kids float on them for fun or you could use one if you needed to transport something down the river. You just hop on the tube, it is a pretty typical way of getting down the river.

ROBSON: So it is not only a hobby, it is also a mode of transport?

MR SAKURA: Yeah, yeah, exactly right. Yah, 20 years ago, there probably weren't even cars right, and if there were, there were very few. People were transporting things down the river on these tubes. Renting them out and floating down the river. I wasn't part of any of that. That was probably a solid four years before I got here. I don't know exactly. The owner of the organic farm says that he invented it but I don't think he did.

I met a couple of Lao guys and they were telling me, 'Oh no, we started the tubing.' Everyone seems to claim that they started it up in one way or another.

I see old signs like my buddy's dad has. He was the first tour guide and he had a sign that said he could take you up 20 km, not just to the organic farm, but the next big town and you could float down from there. It takes hours, oh yeah, all fucking day.

My stepmom said she came here in the 1970s. She is South African and she said it was the place where people came to smoke a bunch of opium and hang. South Africans were one of the few nationalities who were allowed to come to Lao easily back then.

I asked her if she went tubing and she said, 'I don't think it's how you do it now, but there was tubing.'

I am sure they have had it since as long as they have had tyres. Tubing as it is now is probably about 10 years old. I heard about it when I came to Southeast Asia the first time. I had this picture in my head of a really big river and you had these big ropes to pull yourself into these bars but it isn't quite like that.

The first time I went tubing I took a mushroom shake just before I left. Either at Sundowners Bar or a place in town. I hopped in a tuk-tuk thinking, 'I hope I get to the water before these mushrooms kick in,' because I was sitting in the tuk-tuk with a bunch of strangers. I just wanted to float down the river and watch the mountains.

That first time I was floating down I was having such a good time. I stopped off at this bar which used to have a slide. It used to have this little volleyball area and tiered huts, playing loud music mostly to Lao people. I continued floating down and there was another place that sold mushrooms as well as weed. I had another mushroom shake and continued to float.

There wasn't another one until the Final Bar. I ended up there and looked at these mountains. They were just breathing and vibrating. I remember thinking to myself, 'This is sweet,' and so that was my first time tubing. I was trying to avoid the bars as much as I could because I was just tripping balls.

It was pretty epic and I ended up doing that every weekend. After seeing everyone partying and having fun on the swings and everything, I wanted to try the party version too. I went back the next day and took a bottle of whiskey with me. 'Time to get drunk!'

The other way I would tube would be to go climbing. Just take all my gear in a wet pack, tube to the Final Bar and go climbing all afternoon on those limestone karsts. At about four o'clock the party tubers would show up. People would get really fucked up and they would hit the water hard.

I would go tubing every second day. If I wasn't climbing I would be tubing or sometimes I would climb and then tube in the same day. Back then it was only 30,000 kip which was pretty manageable. It wasn't the 115,000 that it is now and you have to be back by six or it's another 20,000. They started increasing the price pretty quickly as it got popular which is understandable. People would bring the tube back at nine or ten o'clock at night. Sometimes they don't even bring the tube back at all so they have to make their money back.

I probably paid the most out of anyone. I went tubing, actually renting a tube, the most out of almost anyone. I used to go every couple of days, either partying or climbing. In my first couple of years here I must have gone at least a hundred times. I am not sure too many people have paid for tubing that many times. Even in the following years I would get a bunch of buddies together and take some mushrooms and float down the river.

Tubing at that time, when there were a few bars on the river; four of them had nice little swings into the water. Imagine a piece of wood attached to a tree with two ropes. You pull it back and hold it in both of your hands, take a running jump and swing into the river. That was pretty fun, you shout, 'Wahey!' or something like that and if you got really good at it, you could do backflips.

Some of them would be really fucking high and you would see some people really eat shit. Sometimes, if someone didn't really understand the physics of it, they would have a terrible fall. You are supposed to lean back and lift your legs up keeping the rope straight. You can propel yourself forward, swing and then jump into the river. They might keep it too slack and as the rope tightens they'd have all their weight on their arms. They'd then tumble; smacking themselves into the water. I would wince and think to myself, 'Ooh, maybe don't try that again.'

It started out as about three or four bars and then as it grew in popularity there were up to about 20. They were all in the same area of the river and then there would be a gap before you reach town which would take about 2 hours to float down. A lot of the time tuk-tuks would pick the drunk people up from the bars. Still, a lot of the time people would do that two hour stretch in the dark and they would be really drunk or drugged up. They might miss town altogether and be stumbling around on the banks of the river.

The problem was: people would just go to the first two or three bars and get stuck there. All of the bar owners downstream would get annoyed. They had opened up a bar on the river thinking they would make a load of money but no one was going to them. Everyone was staying at the first one. There wasn't enough time during the day to get everyone down. People would also just take tuk-tuks back to town after the first couple of bars and not even tube at all. Whenever I went tubing I tried to mobilise the party. I'd have one beer in each bar, but no one ever stuck to it. It was difficult because if you were having a good time in a bar why would you leave?

ROBSON: So for you, there are three types of tubing experience? Partying, climbing and mushrooms?

MR SAKURA: Yeah, pretty much, that would cover the bases.

ROBSON: Would you ever just get a six pack of beer and tube down?

MR SAKURA: Yeah, yeah. Well, it never used to be cans so it was a bit more difficult. Cans of beer really only showed up 3 or 4 years later. Before that, it was only bottles, big bottles, so it wasn't too easy.

ROBSON: When did the whole tubing thing start going crazy?

MR SAKURA: Probably around the time the cans showed up. Hahahahaha. People could take a case of beer with them down the river.

It was around that time that it started to get really heavy and when the other bars opened. There were some rowdy guys working there who did some really stupid things. It was almost like a competition to see who could do the stupidest shit and it was just getting out of hand. They were daring each other to do really unpleasant things. The people of Lao are usually pretty reserved and this was very un-Lao like behaviour. It just escalated really quickly.

There was a two week period of time when a few people died. People say there were 26 dead in that year but I think that number is high. I have never seen an accident that resulted in a death. I have seen a lot of really injured people and I was hitting the river very often. I think a lot of the numbers jumped up because a lot of people report the same incident in different ways. One day someone would say that someone died the previous day and then two days later, again, someone would say, 'Someone died yesterday.' You would never know if it was the same person or not. In any case, a few people had died and there were a couple of important people who died. One was the son of a very important person and that put in motion the clampdown. All the bars were closed down and dismantled.

More recently tubing has restarted but with lots of regulation. Only five bars are allowed to open at any one time, whereas back then, there were twenty or more. They have also tightened up when you can take the tubes out so you cannot rent them after 4 p.m. and you have to bring them back by eight.

There are a lot of river politics. There's a lot more regulation on the river now and all the bar owners and government officials sat down and had a meeting about it. Bars further up the river have to close at certain times to move people along; the first has to shut by three, the next by four and so on. They have to keep people moving and promote actual tubing because otherwise people get a tuk-tuk up to the first bar and a tuk-tuk back. They never actually tube; they just party there and sometimes go across the river. They have taken down all the swings and dismantled the death slide too so there is a lot less jumping into the river which was what was causing all the deaths.

There was this massive slide made of concrete and tiles. It had this little kink at the end of it which if you smashed your head on would give you a fucking concussion and then it would launch you into the river.

This other bar had a zip line called the flying fox and it just stopped abruptly at the end. You would just hit the stopper and then be flung into a moving river at pace. You would see people cartwheel out of control and just smack into the water.

When the river was flowing strongly sometimes you would jump off a swing and by the time you pop up out of the water you're at the next bar. Alternatively, if the river is low, some of these swings that were fine to go on a week ago are, all of a sudden, launching you into a section of river that is only a couple of feet deep. It was really silly. They had to put up signs saying, 'Do not jump here you will die!' but people would just jump over them.

I witnessed someone jumping over a sign that says, 'Don't jump here you will die!'

What the fuck?

In other places, a swing would be meant to go in one direction but if you swung off it in the other direction it would drop onto jagged rocks. In short; alcohol and water don't really mix when you have got lots of young, stupid people full of energy and testosterone. Especially when they are used to western standards of health and safety, and come from countries with lots of regulations. In Europe, America and Australia you wouldn't be allowed to do it as things are made to be safe.

They think to themselves, 'They wouldn't be built if you could hurt yourself.'

In Lao it is slightly different; the bar owners think, 'Well, if people want to jump off that, ok, just be careful.'

ROBSON: This place is crazy for health and safety.

MR SAKURA: Oh yeah, dude, yeah fuck yeah. Take the pool table as an example…

ROBSON: I am going to have a whole chapter just about the pool table.

CHAPTER FOURTEEN

THE POOL TABLE

MR SAKURA: I think it's sturdy but Jesus… people really stomp on it.

ROBSON: I was stomping on it last night.

MR SAKURA: I was just watching it, and I don't really know what the structural integrity of it is. I mean it's put there to fucking play pool on.

ROBSON: Did you buy it or was it already here?

MR SAKURA: It was already here, over on the other side of the bar. It only got moved to the middle of the dance floor when we refurbished the floor. It ended up staying where it is today and then our pool cues just started going missing. We had one little stubby left. Then the balls started going missing too and one day people started dancing on it. At first, I was angry and I was at the back of the bar when I noticed. I ran over telling everyone to get off the fucking table!

ROBSON: You still thought of it as a pool table?

MR SAKURA: It *is* a pool table! I was talking to Joey, the actual owner of Sakura, and he seemed relaxed about it.

He just said, 'Let them, they're having fun. I don't think it will break.'

So I am thinking, 'Well if we're gonna do that, we may as well tear off the felt,' and I wrote 'Caution' on either side. I thought I ought to put a couple more legs on it. People still ask if they can play pool.

When people ask if they can dance on it, I always say, 'Hey, remember it's a pool table. It may not be the safest thing to dance on, just be careful and don't die!'

About a week ago I noticed people were really struggling to get on it, so I came up with the idea of putting some stairs on one side which confirmed it was for dancing. After that, I keep taking a good peek at it to make sure it's ok and it seems to be supported. It's good enough for the time being. I have started to see cracks develop. I don't know how thick it is. One day... one day it is going to collapse and I just hope that it isn't very painful for whoever is on it.

ROBSON: You will basically come in one morning and the first big cracks will appear. I can't imagine it will just drop. It will be pounded and pounded and pounded, probably here in fact, which is where I was dancing last night.

MR SAKURA: Once there's a chunk falling off I will have to haul it to the back, it is really fucking heavy. I think it is made of concrete.

ROBSON: That will be a shame when that happens. It is such an integral part of the bar.

MR SAKURA: Oh yeah, it definitely is. It was a fluke how it happened and people fucking love the pool table man... you should have seen the first time I put the stairs there. That was just a free for all.

ROBSON: Was that the day you saw most people on it?

MR SAKURA: Oh yeah, definitely. I said to Joey, 'The pool table is like really full right now.'

ROBSON: How many people have you seen on it?

MR SAKURA: Oh fuck, I don't know, we'll count tonight.

ROBSON: It is fun when you are on there. Last night, loads of people were on it. I made a video and I was just grabbing people and helping them up on to it. I love the pool table.

I remember being on there thinking, 'Maybe I shouldn't be on here because it's a pool table after all.' Then I saw the stairs so I thought it should be alright.

MR SAKURA: Exactly.

ROBSON: 'I am supposed to be here. That is what they want you to do.' Never once did I think it could be dangerous, like you say, I just got taken away with it in the moment.

MR SAKURA: That is why I put some 'Caution' paint on the cushions, if anything, to stop people tripping over it.

ROBSON: That could be nasty, especially if they then hit their chin.

MR SAKURA: Yeah, but if I take these cushions off it doesn't look like a pool table anymore.

ROBSON: It would just be a platform and then it wouldn't be the same.

MR SAKURA: There's something about that fact that it's a pool table.

CHAPTER FIFTEEN

MR SAKURA ON HIS PARENTS

ROBSON: What do your parents think of you being here?

MR SAKURA: They are very supportive. My mom has been out here twice, my aunt and uncle came out to visit once. My dad doesn't really like flying so he has never been but his wife, the South African, has. She is really supportive. She is the one who came in the 70s.

ROBSON: She sounds like a bit of a hippie if she was here in the 1970s?

MR SAKURA: She really likes to travel. In fact, she is one of the ones who keeps telling me: 'Why do you keep coming back to Canada?'

I respond, 'Well I'm supposed to.'

She always says, 'No! If you're being successful, and I don't mean monetarily, I mean happy. If you are achieving and at peace, not worrying about anything, then keep doing that.' My dad listens to everything she says so he's cool with it. They live on the East Coast of Canada so when I go to Europe it's easy to hop over. It is cheaper to fly to Nova Scotia via London than to fly to Vancouver.

I could stop in Japan and hop across or go via Korea but that's much more expensive. I prefer to spend a month or two hanging out in Europe, meeting up with friends from here which is pretty sweet. It is much cheaper to get around Europe compared to Canada. Canada is large and expensive.

There are no trains and flights are stupid expensive. It doesn't make sense economically a lot of the time, and that's why my stepmom talked me out of going back this summer.

'Why are you going to waste all that money again? Just come home every few years for Christmas. You are where you are supposed to be.'

Like yeah, it's all round pretty positive. I might have some uncles and aunts who are thinking, 'What the fuck is he doing out there?' but my parents are cool.

ROBSON: That is nice. That was one thing I was wondering - maybe you were disconnected from your family, most wouldn't have parents that supportive.

MR SAKURA: My dad was living on the other side of Canada in the first place. Even as a kid, I only saw him three times a year; summer, Easter and Christmas.

ROBSON: Do you Skype?

MR SAKURA: I call my mom regularly; probably once a week or every couple of weeks. I am usually pretty drunk because the time difference works out that way. Midnight, one o'clock here is a really good time for me to call their time.

ROBSON: She has been here a couple of times. What does she think? Has she seen Sakura?

MR SAKURA: She has been to Happy Bar and Sundowners.

ROBSON: What does she think of all the drugs?

MR SAKURA: She has sampled.

ROBSON: Wow! Your mum?

MR SAKURA: Smoking a joint with my mum for the first time, that was kinda awkward. We were all sitting round smoking and she is sitting next to me. I pass it, not to her, but the next person and she reacted.

'What? Is something wrong with me?'

And I was, 'What mum?'

She said, 'Pass that shit over…'

Then she tried some mushrooms and she smoked some opium, well, had some opium tea. The first time people have opiates they get sick, so she didn't like that too much. With the mushrooms, I just let her do that on her own. I didn't want to be around her as I didn't want to trip her out.

You know, doing mushrooms with your son? When the sky opens and you realise the universe is the universe, that might be a bit much, no?

'Mom, I would suggest you do the mushrooms over there… I am just going to let you go for it. You'll love it, just look at the mountains.'

Here is just such a perfect place to do mushrooms. The mushrooms here are really, really friendly, really happy. Giggly, not super-super visual I found.

I used to do a lot of mushrooms; psychedelics are fun, I used to be into it.

ROBSON: You don't do them anymore?

MR SAKURA: Oh no, they take up too much time. To be high for six or eight hours? I usually have shit to do and it's very rare that I have a big chunk of free time.

CHAPTER SIXTEEN

MR SAKURA ON SHOTGUNNING

ROBSON: Did you invent the shotgun?

MR SAKURA: It's a Canadian thing. First time I ever did it, I was 23. I was working in Calgary and I took a knife and just stabbed it right through both sides of the can. It went right through my hand too and, 'Ahhhhh fuck!' that hurt.

ROBSON: Bleeding?

MR SAKURA: Oh yeah, quite well. Yeah, I remember the first time I saw someone bust it with their thumb.

ROBSON: It sounds dangerous?

MR SAKURA: Oh yeah, it is. You take a beer can, turn it on its side. Line it up so the normal drinking hole is pointing up. I take a key and you should be able to feel a small air pocket towards the bottom of the can. Cut out three sides of a square and then push it through. Then you crack open the beer and drink it from where you cut out the square.

ROBSON: I have seen people do it with their teeth.

MR SAKURA: Yeah, you can just crack it with your tooth near the bottom of the can and then open it up. Luckily in England, you have free dental care. I prefer using a key or a knife. I've grown more cautious over the years. I have ended up in hospital a couple of times

ROBSON: Because of shotgunning?

MR SAKURA: Just accidents in general, usually, being careless. Shall we do a shotgun now?

ROBSON: I am working but I guess everyone in your bar seems to work and drink. I guess it's part of the tradition; the shotgun?

MR SAKURA: Yeah. Let's do it. You know the prayer, right? As we say in Vang Vieng:

ALL: Hats off heads,
Hands on hearts,
Praise thee father goose,
Master shotgunner up above!
May these beers go down,
Smoothly,
Efficiently,
With minimal spillage,
And absolutely no after sickness,
Three-us,
Two-us,
Gun-us

SHOTGUN!

ROBSON: Where did the prayer come from?

MR SAKURA: That's another Vang Vieng legend. Thompson. Anton Thompson. From Cornwall. He's the first one to introduce me to the shotgun prayer. It's from Goosin' the Globe. It is a Facebook group. I am selling these T-shirts online to Goosin' the Globe members. They are T-shirts with the prayer but sponsored by Sakura. Actually, Namkong picked up on it and they are going to make us our next round of shirts with a little Namkong logo on it.

ROBSON: You always shotgun Namkong?

MR SAKURA: Yeah it's the cheap one.

ROBSON: Beer Lao is more expensive?

MR SAKURA: Yeah, especially when you go through crates and crates of the stuff. We go through 20 cases of beer every couple of days.

ROBSON: What? The bar or just you and your team?

MR SAKURA: We don't sell too many cans of Namkong to customers, right?

ROBSON: I didn't even know you had beer on the menu; just vodka and balloons.

MR SAKURA: I woke up at eight. I did an eight o'clock shotgun. The guy from Chill Out bar was leaving. I wasn't getting up.

ROBSON: Do you have shotguns for breakfast regularly?

MR SAKURA: When people are leaving or birthdays. It is not by choice, only if you have to.

'Oh really, are you sure you're leaving, come on, stay longer.'

When you had to work hard for business and promote; promoting is always easier when you're a bit drunk. When we went on the river we would create a wizard staff. Have you ever played that?

ROBSON: Like a funnel?

MR SAKURA: We had funnels and shit like that too but we also used to make wizard staffs and wizard swords using cans of beer. Every beer you drink you would tape it to the next one and before too long you have a giant staff of empty beer cans. You would keep drinking the top one until you finished it. Then you would tape a new one on top and crack it open. If you had a staff as tall as you are then you can call yourself a wizard.

A wizard sword was slightly different in that you would tape the first one to your hand and build it from there. After a while, it would get difficult to drink so you have to be innovative. You needed to be skilful pouring into the air and catching it in your mouth or maybe drink from each other's wizard sword.

We had to create innovative ways to drink beer because there was only beer to drink on the river back then. There were no ciders and the alcohol selection was minimal so there was a lot of Beer Lao being drunk.

The other thing we would do is create a necklace out of all of the ring pulls on the cans of beer. You cut a small section on either end and link them together to make a chain. For the best effect, you needed four in each chain link. To make a proper boss chain you would need 248 ring pulls in total. That would take maybe a couple months but you had to drink the 248 yourself. No cheating.

We have a special Sakura T-shirt just for you - the Shotgun O'Clock one. It's got the prayer written on the back

ROBSON: Ah nice.

MR SAKURA: I will hook you up with that and the T-shirt quality is much better; it's one of the good ones.

CHAPTER SEVENTEEN

MR SAKURA ON THE T-SHIRT

ROBSON: This is a book about a T-shirt. The only reason we came here was because of the T-shirt we saw in Luang Prabang. It is why I was fascinated by it because it got me here and you obviously see the T-shirt everywhere. When I started counting T-shirts on my way to Don Det I lost count on the first day because I saw so many.

MR SAKURA: Fucking A!

ROBSON: I love how everyone is just wearing the bloody T-shirts, it's amazing. Where do you get them printed?

MR SAKURA: We have got a couple of people who do them for us. The shop across the street, which is handy and there's also a guy in the Vang Vieng market. The ones from the market are a slightly better quality, you can tell the difference, or at least I can, as they breathe a little better.

ROBSON: Is the one I am wearing right now one of the better quality ones?

MR SAKURA: Yeah you can feel it. They both cost the same.

ROBSON: Can I get a gold one?

MR SAKURA: Gold one… like yellow?

ROBSON: No, gold.

MR SAKURA: Hehehe.

ROBSON: I'd love gold with black writing, or maybe black with gold writing. Maybe I'll have to get it off eBay or get someone to print it back home. I'd quite like a gold one. You keep coming out with different colours. I know the colours are different every night but last night there were four new colours I had never seen before.

MR SAKURA: Yeah, it is actually providing some competition. The guys across the street keep asking me, 'Why are you getting them from the guy in the market? Why don't you get them all from us?'

In response, I'm like, 'Well he's got better quality and he's got different colours.'

'Well we can get better quality and more colours…'

And I'm like, 'Exactly. That's why I'm going to him. Up your game!

I will get 100 off each of you every day if you can both have good quality, but until then, I will take 60 off you and 100 off him.

Competition in the market is good.

ROBSON: How many do you give out a night?

MR SAKURA: At least a hundred. During a lot of the early days we were doing 80 or a bit less, but lately, the last two months, just about 100 a night, sometimes even more. Some nights I will get 180 in and be almost out and be like, 'What the fuck?'

ROBSON: That was probably the night I showed up.

MR SAKURA: No one else in town does them. They all think about it but then they say you aren't really making much money off it, just like all the other businesses.

They ask, 'What are you doing giving away all these T-shirts?' But the T-shirt it is an endorsement, it's like word of mouth.

ROBSON: If you are going to wear the T-shirt it's because you like it, and you like it because the product is good. This is why for me it is genius. It is everywhere. You do know that don't you? You know it is absolutely everywhere?

MR SAKURA: Yeah, it's all I ever hear every fucking day!

'You know your T-shirt is everywhere... in the Philippines, in Indonesia.'

That was true even with Happy Bar too. We were getting that, 'I see your shirt everywhere.' The Sakura ones are just a little more attractive.

ROBSON: They are nice shirts I guess.

MR SAKURA: It's a flower, it's not super effeminate, it's not offensive and it's just... it's nice and... fuck and everyone loves fucking Vang Vieng.

ROBSON: Tell me about the slogan.

MR SAKURA: Ah, that was Chula. Marlon Chula. He got the slogan. 'Drink Triple, See Double, Act Single.' That was about three years ago when we got that.

He is from Vancouver Island, he's a sweet dude. He saw it on some ski resort sign back home or something and just like that I thought, 'Oh yeah that works.'

We were trying all sorts of different shit like, 'Sundowners Bar, shreds the gnar,' it's like a snowboarding term. 'Getting people wasted since... make up some year...' Whatever kind of funny slogan we could think of, we went through a tonne of them. We got in trouble a couple times because our flyers were just ridiculous. Some of them... putting anything on them. Then we started hitting bus stations, flyering there and helping people out.

'Hey, we speak English. You want to stay at a guesthouse? There are some good ones over there,' and pointing them in the general direction of the bar so they were close. There was always the location problem, that was the big thing back then.

Then Happy Bar, same type of vibe. We were really busy every night. Getting busy early in the night. Really good shit and people were saying the same thing, 'Oh yeah, I've seen your T-shirts throughout.'

We had the Nirvana smiley face which I don't think was as appealing. The Sakura shirts have a flower on them. You can wear it wherever and not look like a complete wreck head. With Happy it would have been nice as a souvenir kind of thing but not really wear it all the time.

You should see Pierre. I hope Pierre still wears his one top that a Korean artist did for him. There was this Korean artist who made the one mural in the back. He didn't speak any English but he spoke French. He went to a pretty good art school in France and was just an awesome artist.

Yeah, fuck, this was just after some German carpenters finished refurbishing the bar and I wanted to get as much international flavour in there as possible. So I told the Korean guy you can do as much art as you like. He magic markered up this one guy's top and it looked so awesome. I just don't even know how to explain it but he really spiffed it up and made it look epic. Shit, it's too bad he couldn't wash it.

ROBSON: Could he dry clean it?

MR SAKURA: Maybe. We will see how committed Pierre is tonight… oh look there he is… Pierre!

ROBSON: Speak of the devil…

MR SAKURA: French motherfucker! Pierre where's your top?

ROBSON: I just heard about this special Sakura top.

MR SAKURA: We were just talking about it.

PIERRE: Ah, it's so good. Yeah, I got this one, I will show you definitely. I have it in my bag, so amazing. This crazy Korean guy, yeah, he did it. So good.

MR SAKURA: It *was* so good.
ROBSON: Last night was awesome. I was so excited when I got the white

with the pink trim on.

MR SAKURA: Oh yeah? You found one?

ROBSON: There was only one left and I saw it behind the bar… they were like, 'Do you want the purple one or do you want the grey one?'

I said, 'No, I want that one… oh my God, it's the white with the pink trim!'

MR SAKURA: Oh my God, people, they ask for that one. They will show up even weeks after we had it and ask, 'Do you have the white ones with the pink?'

I would say, 'No but you obviously saw it someplace. We only had 20 made for Canada day.' That one is popular.

ROBSON: That's why I came back. I came back for that one… and the blue one.

MR SAKURA: The blue? It's not so rare…

ROBSON: Maybe not, but you haven't had any in the entire time I've been here. And I've been to Don Det and come back again and you still don't have it.

MR SAKURA: Ha!

ROBSON: Have you been to Don Det? Presumably?

MR SAKURA: Yeah, a few times.

ROBSON: I went to Adam's bar and I met Adam. Have you met Adam?

MR SAKURA: Not sure, maybe.

ROBSON: It was just funny, because he's got T-shirts as well:

'Been There, Don Det' on the front and Adam's Bar on the back.

MR SAKURA: I like that.

ROBSON: I like that too.

MR SAKURA: Nice shirt.

ROBSON: It's a good shirt but he doesn't do a drinks promo. If you want one, you just buy it for 20,000 kip. I don't think you'll get as many of them out there.

MR SAKURA: Exactly. And it was something similar at Angkor What? Bar in Cambodia. You'd get a better price on the shirt if you bought two buckets which were five dollars each. That is ten bucks and then it's another three bucks on top for the T-shirt. That wasn't really the best deal if you want to get a free T-shirt.

ROBSON: You feel good about a free T-shirt, and then you put it on, and you wear it.

MR SAKURA: Oh yeah, definitely.

ROBSON: It's a fucking free T-shirt!

MR SAKURA: Exactly!

CHAPTER EIGHTEEN

MR SAKURA ON DADDY

ROBSON: So… they call you Daddy?

MR SAKURA: Everyone in town calls me Daddy. Behind the bar all the girls are like, 'Daddy, daddy,' I am sure that some people must be thinking, 'What's going on here?'

ROBSON: Where did the nickname come from? Well, not the nickname, the name?

MR SAKURA: It started off at Happy Bar; this girl called Fanny…

ROBSON: Funny name.

MR SAKURA: Great name.

ROBSON: I had to do a double-take when I met her.

MR SAKURA: It gets better, as her boyfriend, in Lao, his name means balls.

ROBSON: No way!

MR SAKURA: Yeah, Fanny and Balls.

ROBSON: That's pretty funny.

MR SAKURA: That is funny. Makes me giggle. Anyway, Fanny just started calling me Daddy one day.

ROBSON: How come?

MR SAKURA: I don't know. I was pretty drunk in those days but for some reason, it ended up sticking. When I came back this time around, Fanny was still here and she told everyone.

'We all call him Daddy.'

At first, I was thinking that it is kind of awkward, and then I guess it fits. I organise their food and their accommodation, I make sure it is all covered.

ROBSON: You are their leader? The head of the family? I guess it fits if you say it is a family.

MR SAKURA: It is a family.

ROBSON: How big is the family? How many kids are we talking about?

MR SAKURA: Just the tourist workers? The ones that call me Daddy? Or if you include just Sakura? Some of the other bars they also call me Daddy too. I think it's eight people right now but a couple of weeks ago it would have been twelve. It's rainy season so people are starting to leave a lot more frequently. Probably up to fifteen when at full swing. We have a lot of promoters and stuff. I usually like to have four westerners behind each bar, then one or two promoters giving out flyers trying to get people in. If it's a quieter night we don't want the party to die.

We need people to go, 'Hey ho, it's balloon time!'

You know, stuff like that to keep the party going and make sure everyone is having a good time. There are certain types of people who are good at that.

ROBSON: It's a skill set, right?

MR SAKURA: Yeah, exactly. Before Daddy, it was T.O. which was short for Tour Operator. It was the same thing. Everyone at Red Bar at the time and Sundowners called me T.O. That was a bit less awkward.

ROBSON: The whole thing fits.

MR SAKURA: If people need to know anything about the town, or Lao, they come to me. Even if I don't know, I can usually find the answer pretty quickly or at least know where to find it. A lot of the shit is actually here, or hidden, you just have to find it.

ROBSON: Connecting people?

MR SAKURA: Yeah, I've recently found that I am a connector. A few different people are leaders, some are followers, I am a connector. My life is about connecting. Even if I'd be in London and throw a party, a lot of people will be there and their connection is me. It is what all these people have in common. Plus, they were both in Vang Vieng separated by two or three years. They are now friends or they are getting married. It is neat finding out people who met in Vang Vieng are getting married.

ROBSON: Sakura weddings?

MR SAKURA: Yeah, yeah, you'll see a load.

ROBSON: Maybe Sakura will breed a load of weddings?

MR SAKURA: Oh yeah, I imagine the one season I really missed. It's too bad, I missed it. Ah, I missed a season.

ROBSON: You gotta make up for it

MR SAKURA: My averages are way down. Slow year, slow year. It was a growing year anyway. I needed to get out for a bit, get a little… [Looks thoughtfully]

ROBSON: That was a theme I wanted to cover today. There's definitely a growing up piece, you definitely seem a different person now to before.

MR SAKURA: Oh yeah, Stephen is different to T.O. and T.O. is different to Daddy. Stephen was my first couple years; climbing and chilling out.

ROBSON: T.O. is Sundowners?

MR SAKURA: Yeah, yeah and that was the crazy time when it was starting to pick up. I met a couple of Dutch guys who were just awesome. They're like my best friends; Bob and Ben, just legends. They gave me the nickname T.O. that just stuck, until I guess when everyone started calling me Daddy at Happy Bar.

ROBSON: Spelt D-A-D-D-Y?

MR SAKURA: Just like your Dad.

ROBSON: Not DAD-E like a gangsta'?

MR SAKURA: Yeah, DAD E DAD! I like it.

ROBSON: Hey, maybe you will become Mr Sakura if that's what you are in the book?

MR SAKURA: Yeah, could be, it's a bit more… a bit more legit. Believable, dealable and makes kind of sense. How about just Stephen?

ROBSON: It's not Stephen! I think we're all agreed, it's definitely not Stephen.

MR SAKURA: Even when I was T.O. I'd introduce myself, 'I'm Stephen.'

People would be, 'Who the fuck is Stephen? I thought your name was Terrence or something.'

ROBSON: Terrence?

MR SAKURA: Funny as fuck.

CHAPTER NINETEEN

THE KOREANS

ROBSON: I find the whole Korean thing here pretty awesome. I love it and I like Korean people and I now have Korean friends. I have even got a Korean fan club of hot girls who follow me around and scream a lot.

MR SAKURA: Nice!

ROBSON: I'm happy about it! But when I was in the bar I noticed there is definitely a divide between the Koreans and the westerners. A couple of people even say, 'Ah, Sakura is great apart from all the Koreans,' and I don't know if you've heard that too?

MR SAKURA: Yeah, I know.

ROBSON: I think it is the Koreans that make the joint!

MR SAKURA: Oh yeah, fuck yeah! If there were no Koreans there'd only be 14 people in there some nights. Without them, there would be no people dancing. They just get straight up there, on the pool table by 8.30 and they are screaming. They turn it into a rave night club. It's more ravey than when I started thanks to them. They like a lot of the hard house which I don't really like so much. Boof, boof, boof, boof. They love it.

ROBSON: What do the Koreans normally drink?

MR SAKURA: Koreans love cocktails. We revamped the cocktail menu once I got here. We had all the names like Blue Lagoon and Vang Vieng Sling. The problem was anytime anyone came to the bar and asked for one it would be made differently every time.

You would have people saying, 'I bought one at the top bar and it's way better than this one. What the fuck?' Stuff like that.

I was trying to get everyone on the same page. One afternoon we took all the stuff on the shelves, put it on the table and said, 'Let's make these drinks with precision. Let's make sure everyone knows how to make a Vang Vieng Sling.' Everyone now knows what is in it, what it looks like and so they are consistent every time.

ROBSON: I thought I would make an effort to connect with the Koreans. I was standing outside with these four Korean girls and Gang Nam Style came on. Of course, they went absolutely mental so I joined in with them. It was probably one of my favourite moments of the whole trip.

MR SAKURA: Nice one!

ROBSON: I literally broke my flip-flop because I danced so hard. You know when it gets unplugged?

MR SAKURA: Oh yeah, I know that one.

ROBSON: I was really annoyed that I had broken my flip-flop, and this old guy, Henry, a real tour guide. He is a Korean but he lives here and he runs tours. He basically picks up my flip-flop and takes his bag out. He pulls out a magic marker! He uses it to thread through the flip-flop. He then held up the flip-flop for me but I took the magic marker instead.

He was left there holding my flip-flop. I got my shirt signed by as many people as possible. I now have a Sakura top that is signed by all the Koreans.

CHAPTER TWENTY

SAKURA WEDDINGS

MR SAKURA: There are a lot of people who meet here and end up getting married. They meet on the river or in the bars and then they'd stay together for years after. Some are even getting married. It's really cool.

ROBSON: That is cool. How many times has that happened?

MR SAKURA: At least three and that's just the Brits. One or two of the Canadians that I know of.

ROBSON: Say about one a year then.

MR SAKURA: Yeah. This place also ruins a lot of relationships, people come here together and leave separately, I wouldn't necessarily say it's the most romantic of places in that respect.

ROBSON: That is also brilliant.

MR SAKURA: A couple who got married last week, Tim and Jenny, they used to work in some bars here. They're in England right now, and Craig and Alannah. Oh, what was that really attractive couple Sara and shit, shit, shit, he's the coolest dude ever… I'll get that. She's so beautiful, and you just want to be jealous, but he's such an awesome dude.

ROBSON: So you don't feel bad.

MR SAKURA: He wasn't super extrovert, he was just a super awesome dude. You're just like, 'Fuck yeah, nice one.' Then there's Johnny Bravo, but I think they just broke up. They were seeing each other for three years. Then the Canadians, I'll get some details.

ROBSON: Did you go to any of the weddings?

MR SAKURA: No, no, I have been out here. I was invited to this most recent one, Tim and Jenny. I was actually hoping I might make it to Europe this summer and go home that way. The way things are going it's gonna have to wait another year. I am sure there will be another… every summer right? Summer is wedding season. Next summer there will be a couple I am sure. I am trying to think of who met three years ago, 'Oh wow, cool, oh you guys are still together.'

ROBSON: You might have a little gap, for the year you were in Canada, if you're saying on average, people get married after three years.

MR SAKURA: David and Sonia, I think they're getting married. Mexican guy, he was just slaying women like crazy then he met Sonia and she stole his heart. Now they're still together in Japan right now, it's like a couple of cool stories.

CHAPTER TWENTY-ONE

THE SAKURA MOUSTACHE

ROBSON: You have a Sakura moustache do you know that?

MR SAKURA: It's ah…

ROBSON: We were in Don Det and we were talking about Sakura, and someone said, 'Sakura? That's a blade isn't it?'

Someone else disagreed, 'No it's a moustache. I think It's a moustache named after a Japanese knife with a blade.'

MR SAKURA: Oh is it?

ROBSON: You didn't know?

MR SAKURA: I got the Sakura moustache?

ROBSON: You have got a Sakura moustache. Did you grow it because of the bar?

MR SAKURA: No, no, no.

ROBSON: So it's just a coincidence that you have a Sakura moustache?

MR SAKURA: Oh yeah, yeah, yeah. I didn't even know it was a Sakura moustache. I just started growing one. I always have some sort of facial hair like beards or moustaches. I always used to rock a handlebar or the handlebars with the chin strap. I had that one going for a while at Happy Bar. It's an easy way for people to identify you.

'Oh look for the guy with the moustache.'

'Which guy with the moustache?'

'The guy with *the* moustache.'

I started growing this one on my 31st birthday last year. I decided I was going to grow a moustache for a year and see what happens. I didn't know if I was going to keep it and make it into a 'dad' moustache by making it really bushy or what I was going to do with it. My buddy, an Australian guy who was out here, came and visited me in Vancouver when I was over there. He had just cut off his moustache. He had a real long twisty one but he was trimming over the lips. I was like, 'You know what, I am just going to not trim it at all until my next birthday.'

He was like, 'Yeah, I kind of wish I could've done that.'

'Yeah, fuck. Why not? I don't give a fuck.'

I don't know. I guess I got some good moustache growing genes. I feel if you can grow a good moustache, you probably should.

ROBSON: There are so few people who can and you can get away with it here?

MR SAKURA: The middle stages are always awkward. Even back home though, I was alright. I was working this job in Vancouver, just before I came here, as a delivery guy. You order your food from a restaurant online and we were the first delivery service that did it. I wore a white shirt with a red bow tie and so the moustache went quite well with the uniform. It worked even with that. The boss, he had a pretty sweet moustache and he was kind of in the middle stages as well when I went in for the interview. We just looked at each other and I said, 'Nice moustache.'

ROBSON: That's probably why you got the job?

MR SAKURA: Oh yeah, yeah, definitely. It's cool enough, it's hipster enough.

A lot of girls are like, 'What the fuck's up with the moustache?'

But then a lot of girls are like, 'That's a sweet moustache I wanna sit on it.'

'Erm, no problem. I think we can figure this out.'

At least one person a day will say, 'That's a sick moustache!' It's kind of a shame, it's always dudes. It's always the guys saying, 'Fucking sweet moustache!'

ROBSON: That is because guys know how hard it is to grow a moustache and girls aren't so turned on by them. They can look a little dodgy like a German porn star from the eighties.

MR SAKURA: Oh yeah, there is always that. On the plus side, girls that like moustaches… *really* like moustaches. Yeah, and they're usually pretty kinky too. I like the girls that are more experimental anyway, the freaky ones, right?

ROBSON: That's a bit of moustache karma.

CHAPTER TWENTY-TWO

MR SAKURA'S MYERS-BRIGGS

ROBSON: Have you heard of Myers-Briggs? Do you know what your profile is?

MR SAKURA: I don't think so.

ROBSON: I discovered this thing at work, about ten years ago, a learning and development piece. You'd like it because of the anthropology aspects. It basically says there are 16 different personality types and you take a questionnaire which matches you to one of them. It costs two grand, but we did it as a team at work so, thankfully, I didn't have to pay for it. When I read mine it was so spot-on I thought, 'Fuck, that is scary how good it is.'

Now I do it to everyone and instead of paying two grand and asking a hundred questions I can do it in four. I can usually tell by talking to people for long enough. I kind of think I know what you are. Do you want to give it a go?

MR SAKURA: Yeah, sure.

ROBSON: Question 1: Extrovert (E) or Introvert (I)? Where do you get your energy from? Do you like to be with people or do you like to be on your own and think things through?

MR SAKURA: I probably get most of my energy from other people. I like to think about stuff and I like to analyse things, but more so people.

ROBSON: Everyone gets a little of both, but it is based on your primary one.

MR SAKURA: Probably go with people… I like conversation and I love hanging out with people. I love chatting and being around people, using other people's energy.

ROBSON: That gives you the E. Question two: Intuitive (N) or Sensing (S)? Are you big picture or detail?

MR SAKURA: Mostly detail. I like things to be done specifically how I want them to be done. You always have to think of the big picture, like with this place, but the detail is how you get there. Coming here there are lots of steps to get it how I want to get it. I did have big picture at first, and then it's how do I get it there which is the detail. I'm definitely more detail. Sometimes I feel a bit OCD. I like everything to match. A black bench next to a brown bench, that drives me nuts.

On most things, I try to be as balanced as I can and try and focus on that… I took a lot of psychology courses which say; if it's not one it's the other. Everyone is just a little more than the other. The more balanced you can be at everything the better you can be as a person. I was being more conscious of trying to balance things out.

ROBSON: Maybe you are dead set in the middle? That would be cool.

MR SAKURA: Balance. That is what I aim for… but I am definitely more detail focussed.

ROBSON: You've got the S. Question 3: Thinking (T) or Feeling (F)? Are you about getting the task done at all costs or maybe you're about how do the people feel about it and seek harmony instead? Maybe it would be better not to do the task at all.

MR SAKURA: Yeah, I would say more harmonious. If I really think something should be done but everyone else is saying, 'I don't know. Stephen, do you really want to do that?' I will ask what everyone thinks.

If I have an idea in mind, I try not to be stuck to it. If I thought, 'Oh wow, this is a great idea, and people are like oh no… fuck ok, fuck I'm wrong. I'm sorry. I don't need to be right.'

ROBSON: That's definitely F. Finally, question 4: Judging (J) or Perceiving (P) Do you write lists and tick things off or do you just wing it?

MR SAKURA: Lists. Daily. Usually, I have to do the orders every day and I have got a stoner's brain. Fuck yeah, I need to make lists, I like to.

ROBSON: You are very organised?

MR SAKURA: I have to be.

ROBSON: That surprises me. You don't come across that you are.

MR SAKURA: I have to. It is something I am a bit anal about.

ROBSON: Is the first thing on your list, 'Make a list' and then you tick it off and you are really happy?

MR SAKURA: Yeah, yeah, could be, yeah. Haha.

ROBSON: So you're an ESFJ.

MR SAKURA: ESFJ!

ROBSON: I have not met an ESFJ this trip. That is quite unique.

MR SAKURA: There you go!

ROBSON: When I get some wifi… if you had wifi I would read you the portrait.

MR SAKURA: That's one thing I could foresee. Especially with internet and word of mouth now compared to three years ago. Social media has taken off. There is a lot more wifi everywhere. We were the first place to have wifi at Sundowners Bar. It brought a lot of people in which is funny because we don't have wifi here. We just keep putting it off.

ROBSON: I don't think you need it. I think people would sit in the corner on their phones and that would ruin the vibe. Now you have to talk to people. This place, it's about partying and meeting people, not about wifi. I think you shouldn't do it.

MR SAKURA: I'll take that little bit of advice and not get wifi.

ROBSON: With the E, you're outgoing so you like conversation, you are very into people; you like people. The F means that you are empathetic to other's needs and want to make sure everyone is happy. The S and the J mean you are really detailed, organised, you like structure, you like planning, you like making things happen. You are quite action oriented

MR SAKURA: Because usually more introverted people could be like that?

ROBSON: Could be. More introverted people would be more focused on detail. Like an ISTJ, for example, would be really into themselves and focus on detail and getting things done. They would be quite one-track minded. They wouldn't talk to anybody they would just do it. People could feel they're not included but they might get things done quicker.

MR SAKURA: I do think I have taken this test though before, it sounds familiar.

ROBSON: I think if you have done psychology and anthropology you probably will have.

MR SAKURA: That was effectively what my degree was all about!

PART THREE

STEPHEN

CHAPTER TWENTY-THREE

STEPHEN, THE SCHOOLBOY

ROBSON: Did you get good grades at school?

MR SAKURA: Oh yeah, straight As mostly. I graduated from high school in Alberta; that's a good place to start. I was vice president of the school's student council. I would have been president but they had a secret meeting in the summer between grade 11 and 12. If you wanted to be president you had to go to this meeting in the summer and no one told me about it. When I got back from the summer break, there was already a president in place so there was no general school wide election but I would have got that for sure.

They didn't want it to be a popularity contest, even though everyone knows student council really is. It's not like I was the coolest guy, or anything close to that, but I had friends in every group of people. So when the election comes along, people would be, 'Ah, I don't want to vote for that guy, he is a fucking asshole to the nerds, or this fit chick because she's a bitch. Stephen, he's cool.'

ROBSON: So you weren't part of the cool kids?

MR SAKURA: I partied with them. I partied with everyone. There were the football players, the hockey players, stoners, the guitar people, gearheads, who put all their money into cars. Like American Graffiti, ever see that movie? Or like Dazed and Confused. My high school was just like the film Dazed and Confused.

Essentially freshmen cruising down the strip, because this was back in a time when there were no cell phones. I was one of the first to get a cell phone back then. Before that, to find out what everyone was doing you'd just cruise the strip asking people where the party was.

Someone would say, 'We know about one, just follow us.' You'd get a long line of cars following each other - the party train. That is why parties would always get out of hand because someone's parents would be out of town for the weekend and all of a sudden this train of cars show up. There were some crazy ones, where front windows got smashed or when people started throwing furniture out of windows. It was fun... high school was fun.

That was also when I started delivering pizza for Panopolis Pizza. I was 16 and had a car so I thought I ought to make some money with it. In the September, I started university in my hometown at the college there. I was a Psychology Major. At uni, everyone started to go to bars instead but we still did lots of cruising around.

By the time summer came rolling by I had moved out of home, started selling a lot of weed and was managing Panopolis. My car kept breaking down so I moved to the kitchen and then they moved me to a supervising position because they were low on supervisors. That summer I partied all the time. I lived with my three buddies and they worked at Panopolis too. We worked, hung out, and partied together every night. I was eating lots of pizza, selling weed and mushrooms and generally just having a good time.

I was managing Panopolis when I started my second year at university. I thought all of this stuff is pretty boring now. I decided I wanted a break and my buddy was moving to Banff for a semester so I thought I'd join him.

In January I went to Banff which is a party town. It's full of tourists, kind of like the Vang Vieng of Canada but with lots of Japanese coming in. It's a ski resort type place, it was pretty cool, and I hung out there for a semester.

ROBSON: I get the impression that you were rocking around Canada; Banff, uni, getting pissed, doing drugs, fucking around...

MR SAKURA: Yeah, exactly. I went back to university to continue a after my second year, I was partying a lot. I think that's when I got a little bored and I was like, 'Ohhhh I want to go out to Southeast Asia. I am gonna go with my buddy.

CHAPTER TWENTY-FOUR

FIRST TRIP TO ASIA AND THE INSPIRATION OF CLIMBING

MR SAKURA: I had my one buddy, Craig Winner, coming out to teach English in Malaysia. He said I should come and meet him in the August, after he finished, and we could travel round Southeast Asia together. It sounded good.

I was flying through Japan and that's when I was talking to my Religious studies Professor. I had taken a couple of Japanese Buddhism classes through him and I told him about my trip. I asked him if there was any way I could get credit for it?

He said I definitely could do and so that worked out really well.

I remember thinking, 'Fuck yeah, sweet.'

This was my first ever trip to Asia. I stopped in Japan for 10 days and then flew on to Malaysia where I met up with Craig. He was living just north of Kuala Lumpur in Salang Jaya. We went out partying in KL to celebrate Malaysian independence day. I was having a really good time and at the end of the night some friendly people beckoned us over and offered us a drink. We had to leave so I smashed it quickly then turned to go. I was walking down the hall, leaving the bar and that was the last thing I remember.

I woke up the next morning in a field. I had no money, no ID, no phone and no idea where I was.

I picked myself up off the ground and saw there was a road nearby. I flagged down a taxi and asked him to take me to Salang Jaya. His English wasn't great and he didn't know where it was. I thought to myself, 'Maybe I'm saying it wrong.' I was walking around aimlessly and had no fucking clue what to do next. I had no way of contacting my buddy and no money to do anything.

At this point I see a black guy walking across the street and thought, 'Fuck he probably speaks English, he is not Asian.'

I start walking towards him and he greets me, 'Dude, you look lost.'

'I don't know what happened,' I explained, 'I woke up in that field over there and I really don't know. Last thing I remember I was in a bar in Kuala Lumpur… I think maybe I was drugged. I need to get back to Jalang Saya or however you say it.'

'Jalang Saya?' He raised an eyebrow. 'You really are lost. That's about 100 kilometres away. That's north of KL. We're South.'

'Ah fuck.'

'Just come with me, let's get you some breakfast, I was going to get some roti anyway. We can figure out what happened.'

'Fuck yeah, that sounds sweet. Thank you so much.'

He was my saviour, I don't know what I would have done if I hadn't bumped into him. We went to a restaurant and sat down, had a chat and ate some roti. It turns out he was a Somalian student living there. Malaysia attracts a lot of students from Muslim countries.

'Why don't you just come back to my apartment, you can get some sleep and have a shower. Don't worry, we will sort you out and find out how to get you back to Jalang Saya.'

'Alright, wicked. I don't know what I did to deserve this but, equally, I don't know what I did to get drugged either.'

I went back to his place and had a nap, hung out with his housemates and listened to some Bob Marley. Someone went out to pick up some burgers and that was sweet.

We tried to figure out what happened to me the previous night. We concluded: I probably got drugged, got into a taxi and didn't have any money to pay. Or I passed out and someone just stole my money and then just left me or something like that.

My new best friend then took me to the bus stop, bought me a ticket and gave me 30 ringgit or about 10 bucks. I don't think he had a lot of money. They were students, five of them all living in one apartment, but they hooked me up.

Amazing. He never asked me for anything ever in return. All he said was, 'Sorry you had this experience. Best of luck.' I never saw him again. He was just awesome. That was sweet. It was a good life experience. Reassuring that people are fucking good.

I ended up getting back to my buddy Craig's apartment at half nine that night.

'What the fuck happened, dude?' he greeted me, 'I just saw you running out of the bar. You just started running through the fucking streets!'

'I don't know man. I woke up in Kapong or Vajong… or something like that.'

'Where the fuck is that?'

'It's far… far away.'

So that was my intro to Asia. And hospitality…

From there we flew into Bangkok and then went to Cambodia. This was another interesting encounter. The first day in Cambodia we show up and eat happy pizza somewhere and before you know it we are already just stoned as fuck. We are wandering around in Sihanoukville, where there wasn't much going on back then. We walk past this little bar with a pool table and some guys are sitting at the window.

'Come in, come in,' they say, 'Come eat some soup with us.'

I look at Craig and then back to them, 'Ok, I am fucking stoned, sounds good.'

We sit down and we're eating this soup. It already looks like they are half way down it, so it's not long before we finish it.

'Oh all the soup is gone, you want to get us another one?'

'Oh yeah sure, one more,' we order another one but don't ask any questions about it. We finish about half of it and finally we are full.

'How much for the soup?' I enquire.

'100 dollars.'

'Er, what?'

'Oh yeah, yeah, you didn't look at the menu? The soup is 100 dollars.'

I am just sitting there with my buddy, stoned, paranoid stoned, with no idea what is going on.

We are just like, 'Fuck man, these guys fucked us,' and like, 'Well I don't have 100 bucks, he doesn't have 100 bucks,' pointing at Winner.

We may as well sit here as long as we can enjoy the company, enjoy the soup, shooting the shit with the guys for the next couple hours.

Then the bar is closing down and we were, 'Oh, about the soup…'

They replied, 'Yeah, don't worry about it.'

I still don't know to this day whether they were trying to fuck us or if they were just fucking with us.

Welcome to Cambodia: check the menu.

From Cambodia, we went up to Northern Thailand, to Chiang Mai and Pai. We were just being straight up regular backpackers. From there we went south to Ton Sai near Krabi. That's the first time I ever went climbing. I was meant to be there for only a couple of days and head to the full moon party in Koh Pha Ngan but after I started climbing I was just like, 'Fuck it, I'm gonna hang around here.'

ROBSON: What prompted you to climb?

MR SAKURA: It started out just as something to do; I was there chilling out on the beach and it looked real cool. I was watching all these other people climbing up and I thought I may as well give it a go. After backpacking for 2-3 months I was feeling a little lethargic and had the urge to do something. I can only sit on the beach for so long, I am more of a mountain, river, lake, kind of guy. So there I was, sitting on the beach thinking I need to do something, 'Fuck it, it was either that or go diving.' The diving wasn't really that good in the area. I would have had to have gone to Koh Tao and I was sick of buses and boats and stuff.

I really fell in love with climbing, fuck yeah that was sweet. I ended up spending two weeks climbing and finished my trip there in Ton Sai. I was supposed to go to a couple more places but basically stopped the trip because of it.

I would climb every two out of three days; you need a rest day, definitely. The first day I couldn't even hold my fork, I remember eating dinner and really struggling but I was just so pumped up. You know that feeling after you go to the gym, really good feeling and a really good pain. After that, I was sold on it.

ROBSON: What do you feel when you climb? Why do you love it so much?

MR SAKURA: It's very relaxing and makes you really focussed, like you can't really be thinking about anything else. If you do, then you're going to fuck up; you're going to fall… and if you fall, falling's not fun because you have to get back up to where you were and it's kind of like… falling sucks… I like to be lazy, I like to do things but I don't like to do the same thing twice.

When you climb you use a carabiner to attach your rope to a bolt and then you keep on climbing until you attach your rope again. If you fall, you only fall as far down as your last bolt. You keep your feet between you and the wall and you drop back down. So every few metres there's a bolt.

It's very relaxing and it makes you feel so good. When you get to the top of something you look at it and are like, 'Fuck, did I do that? Some of this shit looks… there is no way in fuck…'

All you have to do is take it one step at a time and remember to breathe. Every time you go climbing you learn something new. Where I can put my foot and how I can make it easier. There's a real amazing technical aspect to it even if you're not in shape; you can just use your technique. It is very cool.

After my second or third day climbing, I made the realisation that the rope was not actually helping me up. It's only there for safety in case you fall. I am actually getting to the top of all these things by myself, it's kind of, it's really wow. I remember realising that and for me, that was a big part of why I fell so in love with it.

Before that, I was thinking the rope is pulling me up, that it is helping me up, as opposed to helping me from not hurting myself. All the equipment is just for safety, that's the only thing it is there for.

The goal is to do it so you are just climbing up as if it wasn't there, so you shouldn't even use it to rest. You should rest whilst holding onto the wall or standing on a ledge. So when you get to the top you have climbed that all by yourself. You don't use the rope to get up, you just use it for safety and to get back down. You get up to the top, you tie into the anchor and then you call down to the guy below, 'Hey let me down,' and that's pretty fun too.

Some of the people I met were just incredible. They were really, really positive people because they're all like, 'Yeah, just fucking do it, come on,' really supportive all the time. If you fall there is never any embarrassment, it's not even that competitive. I am sure at competitions it is is, but when you are just climbing with friends or just climbing with people around, everyone is in it together.

I would walk around the beach or the bars and ask someone, 'You wanna go climbing?'

'Yeah.'

'I got a rope.'

It's a free activity too which is great or at least very cheap and that meant I could go climbing every day. It's just rope, and I bought some shoes. There's always extra harnesses and ropes because everyone who goes there, goes to climb, and so they all have a shit tonne of gear.

Climbing was awesome, and it really got me into the community of doing shit like that. It only lasted a couple of weeks but it felt so good.

ROBSON: What did you learn about yourself? Do you think you changed in that process; before and after learning to climb?

MR SAKURA: It helped me to build up a lot of trust. I was always pretty trusting, but it is an interesting feeling, relying on someone to save you if you fall. If you do fall, the rope just isn't going to catch you on its own, the belayer, the person at the bottom holding the rope, has to catch you.

There is always someone at the bottom holding you, even if you take hours.

It builds friendships, 'Hey yo, I am Simon.'

'I am Stephen, I am going to trust my life to you.'

I met a few really cool Thai dudes. They taught me some Thai and the Thai way, especially the Southern Thai way, which is really chilled. I smoked a lot of weed with them, like the Thai Rasta guys. I think that part was really special for me as I didn't get that anywhere else. Before that I was just backpacking; a few days here, a few days there. That was the first time I got into the local culture a little bit. Maybe not the culture but hanging out with Thai guys and that was a cool neat little experience.

In that aspect, I just really liked climbing, I was like, 'Ah fuck, I want to climb everything.'

Some of the people that you meet, you think to yourself, 'Holy fuck you're really good - international level.' Amazing climbers come to Ton Sai because it's world class climbing. They are going up shit, like a pane of glass, and I am watching them puzzled thinking, 'This is stupid! How are you doing this? I want to be that good. I want to be...'

It got me aspiring to do more with climbing. It is amazing what you can do with your body. I got more interested in yoga. I had never really pursued yoga up until that point, but it seemed everyone who was good at climbing was also into yoga. You have to know where your centre of balance is.

I also got really into slacklining, which was a tonne of fun. It's like tightrope walking. You take a bit of webbing, like rope, and put it between a couple of trees and you walk across it. That was neat. The trick is: pick a spot and walk towards it. If you look at your feet you will fall. You just look where you are heading, where the rope is tied to the tree and just keep going, one foot in front of the other.

To do that well it really helps knowing where your balance is, where your centre of gravity is. The yoga would help with that. It's all about balance and just getting to know my body a bit better, that was a real good part of that.

I went back to Canada after Ton Sai and wrote about my trip to complete my independent study. I had to write a 20-page essay, and do a presentation on: 'An Introduction to Buddhism: Different Representations of Buddhism in Asia.' It was very broad, but since I stopped in Japan I was able to learn about Zen Buddhism. Japan is amazing, everything is just so beautiful and it all fits into the landscape so well. In Malaysia there are a lot of Chinese Buddhists so I went to a few temples there. In mainland Southeast Asia they are Theravada Buddhists so I got a flavour for that in Angkor and Chiang Mai, which is different to Mahayana - which in itself has different representations anyway.

My professors told me to take some pictures, talk to some monks, just experience it and tell them about it in a essay which I did. I got all the information I needed by the time I had finished in Chiang Mai which gave me a couple of weeks to enjoy the South with no pressure. I could just think about what I was going to write. In reality I ended up partying a little bit and climbing a lot.

As soon as I got back to Canada, the first thing I did was join a climbing club.

CHAPTER TWENTY-FIVE

HOT TUB EPIPHANY

MR SAKURA: I got back to Canada in the December and did another semester. Then what did I do that summer? Work probably. Working. Lots of working in the summers, yeah, bars or mainly pizza places.

My roommate at the time, Charlie Black, was a bar manager and he got me a job at this pretty busy bar. We just smashed it all year long. I learnt a lot about the behind-the-scenes part of how to run a bar. Porting, being at the back of a bar, always making sure everything was stocked. I wasn't bartending but just making sure everything is all right and essentially being the bartenders' bitch.

'Hey we are out of vodka,' or, 'Stephen, can you change the keg?' I would run to the storeroom and do things like that. I really got into it. It was a pretty busy student bar and it was cool to be part of that and with Charlie being the manager, that was cool too. He taught me a lot about bars.

His birthday is the day after mine and he is the complete opposite person to me. He is very negative about a lot of things. He is short and bald which I guess might be what leads him to the negativity, but yeah, he is good shit. Living with him was good times.

I worked with him for about six months and in the June or July, I went back to Banff. I was working in the kitchen at Eddy's, which is a real cool place. It was nice... it was really nice. Banff is really touristy with lots of rich tourists and all the workers, that worked at the bars, would all be hooking each other up.

You would go to one bar, 'Hey buddy, sort me a few drinks and the next time you come to Eddy's I can slide you a meal.' It was a really cool little community, everyone was like 20-something and working in the service industry. It had a very similar vibe to what we have in Vang Vieng right now with all the other workers hooking each other up.

I had planned on spending more time there but after about a month my stepmom rang me up out of the blue.

'Hey, you should come to Calgary! This company I am working for - it's a really good time to get in on it because they are hiring a bunch. You'd be really good at it and I am sure you can jump up real quick.'

So I moved to Calgary and worked for this money transfer service. I started off as a customer service representative but soon after became a trainer. My job was to train the reps on how to answer phone calls to people who just lost a bunch of money. The companies we used to work with were gaming and online poker sites and all that shit. Dodgy tax evasion money and the rest. All the stuff that others wouldn't handle, we would. My company ended up getting into a lot of trouble.

I did that for a year, almost exactly a year and made a bunch of money. I was getting paid pretty well and it was actually a really sweet job; interesting work and I learned a lot about internet stuff and transferring money.

I was driving a Jaguar and doing quite well with the ladies and everything was going well. I remember the moment when I was in the hot tub and I was thinking, 'What's next?' I was renting a house with my buddy so I guess the next step is: buy a house. Then get a girlfriend, get married and have kids, and I saw everything just go by... and I am like, 'I am 23 years old! That's not cool!'

So there I was chilling out in the hot tub, listening to some Bob Marley and then I was like, 'Fuck this shit... I am going back to Asia. I am going back to climbing.'

CHAPTER TWENTY-SIX

THE GERMAN CONNECTION

ROBSON: So that was the moment you shirked off corporate Canada?

MR SAKURA: Yeah, I thought, 'This is neat and everything but I need to get out of here.'

ROBSON: One last hurrah?

MR SAKURA: I didn't know if it was going to be the last one but I just knew I needed another trip. Then right after I left my company, online gaming became illegal in the States, so my company just fell apart. I could've got a good severance package had I stayed around for a couple months but who could have foreseen that. Also, I would have lost my job anyway.

In June I bought a round the world ticket and went to Germany first to watch the World Cup. I stayed with my friend Jurgen and we went to all the fan-fests. I like international football. I am not so bothered about domestic… Champions league I really like. World class. It's awesome.

ROBSON: How did you know Jurgen?

MR SAKURA: When I was 16 I went on a student exchange to Germany. I was sitting in German class one day and it all happened by accident. The only reason I was taking German, instead of French, was because I hated French. In Canada, you have to take French for a number of years. The moment I could drop it, I did. I was doing well in everything else except French. My mum said I had to take another language so I decided on German because my dad is German. Although he can't even speak it.

In one of the classes the teacher talked about a student exchange. I thought, 'Why not?' I didn't really think about it at all until I got this big envelope with all this information about Jurgen, who was coming to stay with me. There were pictures, details of his family, this fucking folder of shit.

'Oh shit! This person is coming in a few weeks!' then it became real. It worked out fine though because he is awesome.

Jurgen came at the beginning of the semester and stayed with me for the first half of the year for three months. He showed up in August and I looked after him, showed him Canada and partied. It was really funny because through the exchange we weren't allowed to drink and you had to abide by both country's laws. Jurgen wouldn't drink when he was in Canada. Even though I took him to all these parties, he was, 'No, I can't, I can't, I can't.'

I was, 'Ah whatever.' He was still fun to have round; I was still getting smashed and got him to drive my car. 'Cool, we always need a designated driver.'

The following March, I went to Germany for the return leg. I was 16 and they were giving me beer for breakfast.

I woke up, 'Here's some roasted rabbit and some wheat beer.'

I was confused, 'But… I am not supposed to drink.'

Their answer to that was, 'You are here to experience German culture, and beer is our culture.'

It's beer and footy. So I got into watching soccer then. First time I ever watched Champions League and learnt how that whole thing worked. Went to a game in Bayern. That whole trip really opened my eyes to what the world was really like. I thought I would be missing out on a bunch of shit at home and at high school. When you are sixteen you think it's a big thing to take off for three months and come back. You worry no one is going to remember you or where your social standing is or whatever the fuck.

I came back and I am thinking, 'What have I missed?' and I didn't miss fuck all. Cool. Yet I had done all this awesome shit.

It was the first time I ever went overseas. That really gave me the idea that you can just take off. I came back after three months in Germany and everyone was just doing the same thing and nothing had really changed. After a couple weeks of people asking me how the trip was, everything went back to normal.

It got me thinking, 'Hey you can just fuck off and no one will really notice.'

Not being able to tell anyone who could relate to it also really helped with that kind of detachment.

'Oh, how was it?' someone would ask.

'It was good.' After that, they don't really care.

Some people would ask me to tell them a story. What do you say?

'You kinda had to be there.'

ROBSON: Like this trip, 'Oh where's your favourite place?'

'Vang Vieng.'

'Why?'

'Oh, I can't even explain why.'

MR SAKURA: Exactly.

'Ooh, just because you got wasted all the time?'

'Ah never mind, "Yes, cause I got wasted," let's move on… what have you been up to?'

ROBSON: Are you still in touch?

MR SAKURA: Yeah.

ROBSON: Has he been here?

MR SAKURA: He hasn't been here, he has been taking law. He is in that whole thing; working in some law office or working his way up to be an international lawyer or something. He's smart, he's a smart motherfucker. He's a good one. Got a good head on his shoulders.

His parents are really traditional, they're actually more worried about me than my own parents. When you asked me what do my parents think, they're cool with it, but Jurgen's parents… I get emails from his mum saying, 'Stephen, you are 30. We think it is now time that you do something with your life.'

ROBSON: 'Do something with your life,' I love that.

MR SAKURA: 'Thanks Gurda, I am good. I will come visit you.' So every time I go to Europe, I take some time to stop there and visit them. My host family. They took care of me. That was a growing… a big growing experience.

ROBSON: That's massive, that really opened your eyes up to the world; the first time you travelled, realising you can take off with no one really missing you.

MR SAKURA: Learning how to be smiling and nod, yeah, smiling and nodding. Learning a lot of body language stuff. I didn't really speak German that well, right? Not many spoke that much English.

ROBSON: Did you connect with German people?

MR SAKURA: Some of them, a bit. It was a pretty small town, I thought I was pretty cool compared to the others.

ROBSON: What town?

MR SAKURA: Oettingen, where the beer comes from. The biggest beer brewery in Germany, but it's not a great beer. Anytime I tell a German guy I lived there, they say, 'Oh the shit beer town.' I thought it was pretty cool anyway.

MR SAKURA: I was immersed in beer culture, I could go anywhere and buy beer, I was partying a lot.

ROBSON: Was Jurgen drinking?

MR SAKURA: Yeah, he didn't drink in Canada but he drank at home.

ROBSON: Did you ask why?

MR SAKURA: It wasn't so much part of the culture in Canada. He was going to Canada to learn English and to go to a different country. Learning a different culture… smart motherfucker.

ROBSON: That's why he's gone to law school…

MR SAKURA: And I own a bar.

ROBSON: Did you teach him how to shotgun?

MR SAKURA: I didn't know shotgun at that time, that wasn't until Calgary. That was the first time I did it in Calgary, working for the money transfer service and I stabbed my hand.

ROBSON: Did you have to go to hospital?

MR SAKURA: No, no, the alcohol, kind of sorted it out, I just tied it up.

ROBSON: That was one of your life's pivotal moments then, that taught you how you could just take off. That is also the reason you are in Germany; to see Jurgen all those years later?

MR SAKURA: Yeah, I stayed with him and we did a little Euro-trip; France and Italy, Slovakia, Slovenia and Prague. That was sweet. After that, I went back to Ton Sai on my round the world ticket.

I thought I would do Europe and then get back to climbing. So went straight there and met a bunch of cool people. I was there for probably three months; Europe in June, July, August and at the end of August I went to Ton Sai. I stayed there for two months because then I went to Korea. I went to Korea in the October… October I go to Korea and I teach Tae Kwon Do.

CHAPTER TWENTY-SEVEN

KOREA CALLING

MR SAKURA: Yeah. I was in Ton Sai and got a call from my buddy Craig Winner who I went to Asia with the first time.

'Hey, do you want to come to Korea and teach Tae Kwon Do?'

'Uh, I don't know Tae Kwon Do.'

'Yah but you don't need to. There's this other Canadian named Stephen who is just leaving. He was teaching Tae Kwon Do and they just needed someone to take his place. Just lead the stretches… well pretty much just get here and they will tell you what to do. You don't need to know Tae Kwon Do.'

So I got there and I lead the stretches. I did judo when I was a kid so how hard could it be? There was this English language martial arts school and the gimmick was that it was taught in English. I used to stand in front of the class and say, 'High kicks, one, two, three…'

I was just reading off a list. I didn't even need to know Korean or anything. After six hours of classes a day they would teach me a little Tae Kwon Do. They wanted me to do it properly.

By the time December came around I am thinking to myself, 'This is fucked up,' and I wasn't smoking any weed because it's really expensive in Korea. This was the first time I hadn't smoked for a long period of time, like in ten years.

One day I was yelling at some kid because he wasn't kicking properly.

The sensei was telling me, 'You have to be more strict,' and I am a real chill guy. He wanted me to yell more at this kid for not kicking properly.

I was thinking, 'What the fuck, I don't even know how to kick properly! This is just weird. Why am I doing this? This is so fucked up.'

So I went home for Christmas and surprised my mom.

ROBSON: Did you learn a lot about Korean culture in those four months?

MR SAKURA: I learned a lot, yeah. It was pretty boring because I had a lot of free time. When I finished the lessons I had nothing to do and they didn't really start till noonish, 1 till 7 p.m. so I had a lot of free time on my hands.

ROBSON: What did you do with the rest of your time?

MR SAKURA: I would go out and drink in Korean bars and learn Korean drinking games which, coincidentally, I played with a couple of Korean guys last night. Titanic. You get a glass of beer and a little shot glass floating in it. You use Soju, a Korean alcoholic liqueur, and you pour it in. You have to pour at least a drop and then it is the next person's go to pour a little bit more Soju in. Whoever sinks it, has to drink the whole thing. It's fun. Real fun. And I fucking rule at it.

I was beating all these Korean guys at their own game. The key is to pour in so much and not sink it so you fuck up the next guy and it doesn't get back round to you.

I was hanging around with my buddy Winner, in Korea, and we were going to all these different Korean bars and Korean BBQs. He had a couple of Korean friends and the people I worked with would take us out to cafes or stuff. It was really cool.

After I left town when I finished doing the Tae Kwon Do thing, I went up to Seoul for ten days and did the Tourist thing. I went couchsurfing. I did couch surfing throughout Europe, Canada and America. That's a real cool way to go travelling. I love that site. Anyway, I was couch surfing in Seoul. Hanging out there was really cool because I wasn't pressured to do anything. I was just hanging out thinking, 'This is a real cool place.' It was the beginning of December, and they have late autumns there. The leaves were changing. It was cool. Chilly, but beautiful.

ROBSON: Do you think you learned anything from that trip that is helping you now with the Koreans here?

MR SAKURA: I think a little bit. There are a couple of things I wanted to do that I haven't quite done yet. They always have food with their drinks which we don't have. I don't want to have food because then it will fuck up some of our relationships with restaurants, and I don't want to compete with them. The Koreans, though, always have food with their drinking; a plate of fruit or whatever.

When you go into clubs there, it is really funny because they would play heavy tunes and then a really slow song. That would break it up and everyone would go back and sit at their table and drink and eat for a while. They would eat their fruit and then the hard dance songs would come back on and they'd go up and dance some more.

Fill up and empty, fill up and empty. That is what's kind of cool about here at Sakura, we just keep it going… woo, wooooo, woooooooo. It builds them up right.

They really like sitting down in their own booths. Booths are everywhere; in clubs, restaurants. You go with the people you are drinking with and traditionally you don't really hang out with too many other people. I got into trouble a couple times in Korea with that. I'd be pretty drunk and just sit down with somebody at their booth and be, 'Hey guys, what's up?'

They're like, 'What the fuck is this guy doing at our table?'

That was just what I would do back home, anywhere in Canada, I would sit down in anyone's booth.

'Hey guys, what you doing?' chat for a bit and then, 'Ah cool, great, alright, wicked see you later.' That didn't work so well in Korea.

We also found out that they like to drink. I mean really like to drink. When the guys get off work they go straight to the bar or to the Korean BBQ with their work mates and stay there until midnight or even later. They don't spend a lot of time at home.

Before you get married, the Korean guys do not know how to cook at all and it's a foodie kind of culture. If you are a bachelor you eat ramen like instant noodles all the time. That's why they have so many brands, and that's why the best instant noodle packets are Korean. They have three different pouches of all sorts of shit to put on them. Mmm, so good.

I was eating some noodles and when the Koreans walked by I am like,

'Mmm, oh yeah, I love this shit.'

When I tell Korean people I have been there, they love it. Especially when I tell them I was a Tae Kwon Do instructor in Korea. I am pretty sure every Korean, every single Korean, at one point or other in their lives has to take Tae Kwon Do.

ROBSON: Were you allowed to teach? I thought you had to have a black belt to teach?

MR SAKURA: Legally, I needed a black belt but my sensei buddy was high up in the thing and he managed to just get me one. I am a black belt. I am a fully licensed, or at least I was a fully licensed for however long it was valid for… fully licensed black belt Tae Kwon Do instructor.

ROBSON: That's awesome. I did Tae Kwon Do for two years.

MR SAKURA: I did it for two months.

ROBSON: I used to go five times a week. I only got to blue.

MR SAKURA: Hahaha!

ROBSON: You just fucking bought a black belt!

MR SAKURA: No, no, not even that. They paid me.

ROBSON: That's ridiculous. In fact you don't even say sensei. That's karate!

MR SAKURA: They asked me 'What's the English word for this?'

I said, 'We always say sensei.'

They were, 'Noooo, that's Japanese!'

They hate the Japanese so much. They hate it when you mix up any Korean words with Japanese.

I would say, 'They are so similar!'

They would say, 'No!'

'Ok, so what is cheers in Korean?'

'Kampe.'

'What is it in Japanese?'

'Kampai.'

'See. Not so different! Soju, it's just like sake.'

'It's nothing like sake.'

'It kinda is. You drink it cold, they drink it warm. It's the same alcohol content, it's made out of rice.'

'They use metal chopsticks, not wooden ones…'

I learned a lot about the history. The Japanese came in and fucked Korea a few times. It would be, 'Here we go again.' They would just come in, rape and pillage, on their way to China,

'I call it the sea of Japan.'

'No, it's the sea of Korea.'

You know the body of water between Japan and Korea, they both call it different things. Everything that has anything to do with Japan, they say, 'No, it's Korean.'

That experience definitely gives me something to talk to the Koreans about. This one guy, Ed, who used to work at Sundowners for some time, always had something to say about anywhere in the world where someone was from.

'You're from Stuttgart? Oh yeah, you know that one bar down the street from there, I've been there.' He would have something in common with everyone and immediately everyone just loves the shit out of him.

'This guy knows where I am from. It's cool.' That is something I try to use as much as I can.

ROBSON: That's interesting; all these experiences when you tie them together. It's really random getting the call to go to Korea and then here, one of your best customers is Korean.

MR SAKURA: Yeah, why the fuck would I go to Korea? I wasn't planning on going to Korea but I had a round the world ticket so it didn't cost me anything anyway. It was just continuing in the same direction so I thought I may as well go there.

ROBSON: And now, nine years later, how invaluable that experience is…

MR SAKURA: Oh yeah for sure, all this shit, nothing is linear.

ROBSON: Which is just another coincidence… or fate?

MR SAKURA: Yeah, exactly.

CHAPTER TWENTY-EIGHT

DISCOVERING LAO

ROBSON: You went back to Canada after another life changing trip. Life changing experience number two? Or three? First Germany, then Asia the first time and then this round the world via Korea.

MR SAKURA: I had come back for Christmas and I was like, 'Ahhh I don't know what to do!' I was just working, doing Two Dudes Pizza full time. I was in the kitchen. They hired me as a manager to start off with. I was also working at an Italian restaurant at the same time and a bar too. Just working, working, working. Then by February, I was like, 'Nah, this actually sucks,' and I went straight back to Ton Sai.

ROBSON: With no intention of coming back?

MR SAKURA: My return flight was booked for June. I was originally going away for a year anyway, June to June, which got fucked up with the whole Korea thing. I went home for Christmas and worked as much as I could for a month or two. That was one thing Two Dudes was so good for, I could get some money real quick. They were like, 'Come back whenever.'

I went straight back to Ton Sai but it wasn't the same vibe as it was the previous year. All the bars on Ton Sai were on squatted land. There is a path that goes straight through this land and all the bars, guesthouses and restaurants that were on the one side of the path were not allowed to be there. There is a law in Thailand that if you squat on land for 10 years then the squatters gain the rights to the land. The landowners didn't want to give their land away so it got quite aggressive.

There was a lot of bad vibes going round and some of the Thai guys I was hanging out with before were a bit bored of the novelty of this westerner. I wasn't too sure where I stood with them because since I got back it was different. Previously, when I was hanging out at Pee's climbing school, I felt part of the family but when I went back everyone was on edge. I don't know if it was because everyone was losing their business but they weren't the same.

So I was at a loss on what to do. I wanted to climb and had my climbing gear. Someone told me I had two options; either go to Chiang Mai or go to Vang Vieng. I had been to Chiang Mai before and I had always wanted to visit Lao so I opted to come here instead. I had heard of tubing before but that wasn't really part of my reason for coming.

I was with a girl then and suggested we go to Lao together. 'Sweet. Sounds good. Let's go up there, and go climbing.' That was about it. I also met up again with Craig Winner. He's one of my bestest buddies.

ROBSON: Is he like the angel and you are the devil?

MR SAKURA: Yeah.

ROBSON: He is your good conscience?

MR SAKURA: Yeah, he's always doing what I think I should be doing. If I had a clone, he'd be doing what Craig is doing. He is going out to China, to make some money and get more teaching experience, and then he is going to come back and help set up a school. That's in a year or something like that.

ROBSON: Does he party as hard as you?

MR SAKURA: No. He smokes a lot of weed. He is a stoner. He is one of those people who can't operate unless he has had a joint. I was supposed to meet him in Luang Prabang. This girl and I stop off here in Vang Vieng on the way and we chill out here. I don't think we even spent a night in Vientiane, and if we did, it's not memorable, but that's just Vientiane for you.

ROBSON: Hashtag Vientiane.

MR SAKURA: I know, right? Then we came here. I knew Craig was in Luang Prabang and planned to meet him there. I spent a week here climbing, smoking weed and chilling out. I went for plenty of walks exploring and I think one of my walks ended up at Sundowners Bar, as it was the farthest away, on the other end of town. It was a real chill spot. Great place to smoke some weed and whatever. At that point I hadn't even met Joey, I just spent a lot of time in the hammocks with a joint. I might have met him or perhaps his brother sold me something. I hung out there for a bit.

Then I went up to Luang Prabang with the girl and met up with Craig. We spent three days there. It was all so very nice, but I was like, 'Oh yeah, yah, this is nice whatever. I am gonna go back to Vang Vieng and climb.'

Craig agreed, 'Oh I will come down with you for a couple days and maybe we will continue travelling together,' Craig wasn't into climbing at all. His plan was to continue travelling.

'Oh I am going to stay in Vang Vieng and climb for the rest of the month or at least until the end of my visa.'

He responded, 'Ok, I will come with you for a while and then continue travelling.'

ROBSON: What about the girl?

MR SAKURA: The girl? I think she took off to Chiang Mai from Luang Prabang. Something like that. She is average anyway. Nice enough girl. She liked climbing.

Craig and I came down here and he said, 'We should probably find a place where we can chill and smoke.'

That is when I remembered, 'Oh yeah, I know this cool spot where we can go.'

So we went down to Sundowners and that was when we met Joey for the first time.

Even then, he would ask us, 'Oh you wanna help make a flyer?'

'Yeah ok, we can do,' Craig and I made a little flyer, like for a full moon party or something like that. At that point, Joey was already interested in getting his bar busy. I wasn't really thinking about it and neither was Craig.

ROBSON: You just wanted to get stoned, go climbing and hang out with Craig?

MR SAKURA: Yeah, yeah exactly. Smoke some weed. We thought it was pretty funny at the time. We were there every day, smoking weed and going tubing. The first time I went tubing was with Craig. We got to see the chilling side of Vang Vieng. First time, I was climbing for a week with that chick and then second time just chilling. Then when Craig left I went back to climbing. It was climbing, climbing, climbing, hanging out at Sundowners. I got to know Joey a little better. We chatted.

I was here for a month and I just fell in love with the place. Sakura was already here but it was run by a Chinese family. It was kinda, yeah, I don't know. They had a couple of western staff, they were doing alright. Sakura was here and they all left or something, they just weren't trying anymore. It turned into a place where you watched movies during the day. The whole back area was set up for lounging with little triangle cushions and a movie screen. It was a chilled bar. Very chill.

Back then, the road in front of Sakura wasn't paved. There were probably only two paved roads in the whole of the town at that time. It was not really too developed yet. There had been a crackdown even a few years before that. It was really quiet, I was just climbing, smoking a bunch of weed and getting to know a lot of the Lao people in town and some of the tour operators.

I'd go tubing, just throw my climbing shoes in my dry bag and get a tube, float down, stop at a bar and chill for a little bit. I would eventually end up at the Final Bar. Have you been tubing yet?

ROBSON: Yes.

MR SAKURA: Yeah, so at the end, by the big turn right next to Final Bar, there's some really good climbing there. I would get off and if there were some people climbing, I'd be like, 'Hey can I hop on with you?'

That was really the Vang Vieng that I fell in love with. Right at the beginning. Then any day I wasn't climbing I'd hang out at Sundowners. That was the first month that I was here.

Right close to the time when my visa was about to run out I was writing in my journal.

'What am I going to do next? Am I going to go to another country? I'm really digging it here… what should I do?'

That's when I got chatting to Joey and he said to me, 'You're here all the time, what are you doing?

'I don't know, I just came here to climb.'

'Where are you going next?'

'I don't know, I was actually just asking myself that in my journal.'

'Do you want to help us get busy? You are our only customer, pretty much every day. My brother's wife, wants you to work for us. Do you want to do this kind of thing? We will pay for your accommodation and your visas and you can smoke as much weed and drink as much beer as you want.'

'Fuck yeah,' I had already made up my mind. I was like, 'You had me at 'Do you want to help?''

I had until June for my return flight and I had already been to Cambodia so I didn't need to go back there. Vietnam is a pain in the ass to get to and I didn't want to sit on a bus for 30 hours.

'Yeah fuck, I can stay here and help get this place busy. I can work in a bar. I have done bar work before.'

That's how it started.

PART FOUR

T.O.

CHAPTER TWENTY-NINE

SUNDOWNERS BAR

MR SAKURA: I said to Joey, 'We can have some sort of party or whatever. I've got some ideas. We can make some flyers. We can compete with Happy Bar,' which was next door and was pretty busy back then.

I made a load of different flyers and I got a little creative with it. I was never really artistic at school but I got into it.

'I know I can do something, I've seen a lot of flyers before, I know what they should say on them.'

We put maps on them which I don't think anyone had done before because it was a hard bar to find. Then we gave it a go with the T-shirts. I remembered seeing them in Siem Reap at Angkor What? Bar. Buy two buckets and get a T-shirt. I thought, 'That's pretty smart, I'm gonna use it.'

We started doing T-shirts, I always knew that was going to be a good idea. They started off very, very basic but I always knew T-shirts were a good one. We didn't really figure out to give them out for free until later.

Sundowners started to get steadily busy. It was gradual at first but for some reason, all the western staff that were in town left and I was the only one around for the next bunch of months and that made it pretty easy to get people to Sundowners Bar.

When I first started it was huts with hammocks. It was chilled. There wasn't really a space to party. But people took weed, mushrooms and opium, so it was a bit more like that type of vibe. We'd get a projector and play funky movies like Alice in Wonderland and fucked up shit like that. People would be rocking the mushrooms and having a good time like that. We'd have fires at night and be open late night, midnight to three or four, and be the later party bar. That was kind of the beginning of how it happened.

ROBSON: So you never set out to create a whole revolution?

MR SAKURA: Oh no, no, no, hell no…

ROBSON: Or even to set up a bar even?

MR SAKURA: No, oh no, I was here to climb.

ROBSON: Climb and smoke weed?

MR SAKURA: Climb, smoke weed, yeah and chill out. Then it just kind of grew from there. Once we started getting a little bit busy Joey's family were so pumped up, I was feeling really good about it, 'Fuck wicked.'

ROBSON: They'd never seen anything like it?

MR SAKURA: It was an empty bar. The first night we made some good money, it was just, 'What the… wow!'

So anytime we bought something, like we went out, and bought the family a new home, they would say, 'Stephen built this house.' They really, really liked what I was doing… Joey's family, they treat me like their brother or son or whatever.

ROBSON: Because you helped them?

MR SAKURA: I helped them get the bar busy. So from March to June, four months, maybe three and a half months I was here.

ROBSON: Were you still thinking at that point you would go home? Stay a couple of months here, help out and go home?

MR SAKURA: Oh yeah, June was when I was supposed to be going home but from May things were really starting to pick up. By the time June came around I was just, 'Fuck, I don't want to go home.' I needed to for some reason, I don't know… a wedding.

ROBSON: I was about to say was it a wedding?

MR SAKURA: Yah, two weddings, and I was a part of the wedding party for both of them. Those were both in July and August that summer. I had to go back and I was sad. I was thinking, 'What the fuck? This is really fun. I don't want to go.' I was part of the family pretty much. They loved me, I loved them. It was working out real well.

I was like, 'Fuck what am I going to do?'

So I called up one of my buddies from back home and was like, 'Hey you should really come out here and just make sure it keeps going while I go home and save money so I can come back.'

I have known Dean since I was five or six. At that point in time he was just hanging out in Canada and he had to leave for legal reasons, anyway, so I told him, 'I know a good place, come here and help the family.'

He never really said yes or no, he was just like, 'Oh alright, I don't really have anything else to do,' and that was the last I heard of it.

Three weeks later, he shows up at Sundowners with a tiny little backpack and asks, 'Hey is Stephen here?'

I looked over, 'What the fu…?'

He was completely nonchalant, 'Oh hey, what's up?'

He came out here then and has pretty much stayed here ever since. He is married to a local girl and has got a kid now.

They had this really nice leaving ceremony for me at Sundowners called a Baci. It is a Lao tradition and is a good luck ceremony you have when someone moves house or leaves on a trip and stuff like that. It was real touching… but then, 'Ah fuck, I gotta go home.'

CHAPTER THIRTY

HOW DO I GET BACK TO LAO?

MR SAKURA: I went back to Canada in the June of that year, went to the weddings and started another semester at uni. I moved into a house with Charlie Black and was working as much as I could to put some money together whilst doing all these classes for my Psychology degree. The whole time I am thinking, 'How do I get back to Lao?'

By the end of the semester, I thought, 'Wait a minute, maybe I could do some more independent studies.' That was when I figured out, if I switched from Psychology to Religious Studies and Anthropology, I could get course credits by coming to Asia to do independent research papers. Essentially I could spend the whole second semester of each year out here and just write essays at the end of it.

ROBSON: That allowed you to maximise your time here which is incredibly clever.

MR SAKURA: It's definitely out there. I don't think anyone else thought of it but my professors were supportive of me doing it.

ROBSON: They didn't think that you were cheating or had found a loophole and were exploiting it?

MR SAKURA: No, no they were fine. Every time I said to the director that I wanted to do another independent study he was fine with it.

He'd ask, 'What do you want to write about?'

'Well like, I don't know.'

He just said to me, 'Wherever you go, write an essay on what you did, what you saw, who you talked to, just tell us everything.' So I pretty much got my degree by coming out here.

ROBSON: That's awesome!

MR SAKURA: Yeah! I was an anthropologist doing anthropological research and I was writing essays on life in Lao. The professors were super helpful getting me set up with that. They said every experience I have is unique and is working towards something.

They told me, 'If you write an essay and someone reads about it, they can understand what life is like in Lao. You can get credit for it. Go for it!'

I was getting student loans to buy my plane tickets and made that a habit. The first semester, September to December, all I could think was, 'Gotta get back to Lao, gotta get back to Lao, gotta get back to Lao.'

When flight prices went down, usually on January 6th, I'd pop back out here and help with the bar.

ROBSON: I think that that is pretty interesting. You did your year out, you are supposed to grow up, supposed to return to society, to go back to uni and settle down. Instead, you've got in the back of your mind, 'I gotta go back to Lao, I gotta go back to Lao,' so what is it that is drawing you back? Why do you have to get back to Lao so desperately?

MR SAKURA: The family. That was it. It was all Joey, the family and Lao people. I had to get back and help them. We got a good start and we really started clicking just as I was leaving. I thought, 'I could actually start something here, get something going. Maybe there's something in the future for me,' that kind of thing. They treated me so well and everything, I felt it would have been rude to not go back.

ROBSON: Because he needed you?

MR SAKURA: Because he didn't know how to do these flyers or get a bar off the ground.

ROBSON: You're like, 'I know this shit, I have been to bars all my life, I know hospitality, this guy needs my help.'

MR SAKURA: Yeah, 'I wanna go back and help him.'

ROBSON: So it wasn't about being Mr Sakura or thinking 'I want to be king of the bar,' or Daddy, which is why you don't like this attention. Your mindset is, 'I am just here to help out my friend.'

MR SAKURA: Yes, exactly!

ROBSON: And if you have to use that personality, that persona, to make the bar a little better than you will, but primarily, you are here to help this guy; it's not about the money, the fame the fortune. It isn't even about Sakura?

MR SAKURA: I am just so stoked that he is so stoked. It's going so well. When he got back from the Philippines last night I threw a tonne of money at him.

ROBSON: You did? Because you had so many good takings?

MR SAKURA: Yeah.

ROBSON: That's because I showed up and started collecting T-shirts, that's *my* money you threw at him!

MR SAKURA: Oh fuck, haha, you should have seen his face, it was priceless… like we fucking actually did it this year… touch wood obviously. Actually, doing everything right. Getting it done. Going on vacations.

ROBSON: He hasn't been on vacation before?

MR SAKURA: I took him down to Kuala Lumpur at the end of the season. We did alright that year and it was pretty much my attitude to help the family and really do this. Hell, if I can get course credit for my degree whilst I am doing it, why not? I'll use the student loan to pay for my flight; it sure beats hanging out at home, like, it's fucking windy as fuck there. I hate the wind and it's really windy in March.

ROBSON: Which is high season here?

MR SAKURA: Right! At that point I was figuring it out; I can live both lives. I can do both things.

ROBSON: This is the thing right, everyone else, they have their wonderful travel experience and then go back to the real world and dream of Lao, dream of Asia and dream of these wonderful places. You are always there back home in reality. This is the point in time that you split, you suddenly go from being, 'That was a backpacking adventure,' to, 'Now I am going to have two lives; half a life in Canada saving money and doing the compulsory courses, merged with my other half of life with this family in Lao where I am still getting credit and I am having fun doing it.'

MR SAKURA: Yeah, yeah, exactly.

ROBSON: And the shotgunning and the getting laid 200 times and the 'Drink Triple'…?

MR SAKURA: That is all part of it. Hell, if I can have fun whilst I'm helping the family, I might as well.

Dean came back to Canada at the end of that year for a short time and he said to me, 'This is stupid, let's both go back to Asia. It's awesome there. Let's do it.'

ROBSON: He must have had his own little adventure for that six months.

MR SAKURA: Oh yeah, he was learning Lao like crazy. He's really smart with languages. He is good with people too, he has his own thing going. He fell in love with a Lao girl out here. She has a clothing shop.

ROBSON: Is he still here?

MR SAKURA: Yes, he spends a lot of time with his family. He is a character and a half too yeah, ha, ha, ha.

ROBSON: This book is mushrooming with so many back stories and side stories.

MR SAKURA: So maybe if you get the first book done, then you can start living here, and then you can write other books. If you get the base one complete and then people will like a character and then fucking do a book on that character. It will be just as interesting. Everyone has got their story of how they got here and what they were doing before and why they came back… and everyone comes back. Everyone that stays comes back. That or they went off to get married together. There are so many little stories we haven't even got to yet.

CHAPTER THIRTY-ONE

DESTINATION LAO — DOUBLE LIFE

MR SAKURA: I'm heading back to Lao so fly into Hong Kong and go overland to Vang Vieng. I saw the weather in Double V was over 30 degrees and it was really hot in Hong Kong too. I was like, 'Sweet!' I got rid of my jacket and everything and then there was this huge mountain pass to go through. It was January in China and it was two degrees, almost snowing. I was sitting there shivering in Nanjing in China, 'Fuck, I am an idiot.'

PIERRE: I was riding a bike in the mountains in Vietnam. I almost got sunburn at the bottom of the mountain and at the top, I was freezing.

MR SAKURA: Yeah, elevation, who knew?

I came back here and Joey was real excited to see me. He and his family were asking me, 'You gonna be serious? You gonna come back every year?'

'Yeah, fuck, if I can work it, I think so. If I am getting course credit for this. Cha-ching.' So I was here until July that year.

Sundowners was a collection of huts that were temporary. In rainy season the island would go underwater and it would be a good time for me to go home anyway. It worked out great. Ok, rainy season, time for Stephen to go home, and I'd come back just as high season was about to pick up. I'd usually be a couple weeks late for it because of the January 6th price drop thing, but that meant I got to spend Christmas and New Year's at home.

That year we built on what we had done previously. I would get it going and it picked up some momentum. More people were coming and I met cooler and cooler people who were all willing to help. I started drawing a staff base as other travellers came and wanted to get the bar busy too. Every month it got a little bit bigger.

That is when I started to do a lot of tours around the country. I gained a few regulars who would also want to see more of Lao. I would know a bit more language as well so, 'Let's rent a minivan and go to Champasak… or go to wherever the fuck.' I had some friends from home come out to visit too. VV was still picking up steam but not 'party-party' yet. We were selling a lot of alternative things at that point.

ROBSON: That was how you were funding yourself?

MR SAKURA: No, that was what we were selling at Sundowners. Happy Bar was busy during these years and we were the alternative bar next door.

Back then, I was still tubing, going down the river, still doing a lot of climbing. This was the year I started helping with some of the tour groups; taking people out climbing, kayaking or helping out other tourist businesses as well as working at the bar.

I was 25, smoking, climbing, kayaking. It was a dream life. The bar was slowly getting a little busier. I had Dean working with me, and there was a couple of us working at the bar and getting known as the only foreigners working in town in bars. We were slowly getting a bit of a reputation, these Canadians, and that was making Sundowners busy.

We were working at the bar, still tubing down the whole river. That whole tubing bars on the river was still fledgeling, they hadn't built any bars where the tuk-tuks would drop you off yet. They did that at the end of that season. That is when Bar Uno showed up which was the first proper tubing bar, right where you got dropped off, right where Life Bar is now.

That was also when shotgunning started; when the beer cans showed up. Prior to that, there were no beer cans. I think they were trying to limit the bottles on the river because of the amount of glass. That was the year Dean and I both sliced our feet. He cut himself on a bottle of glass and me on a sharp bit of rock.

ROBSON: Whilst you were tubing?

MR SAKURA: Yeah, it was my last day of this year so it was the July of the season and it sucked. I was supposed to have a nice big going away party. I was tuber number one that day. I wanted to be the first tuber and get the number one on my hand. At that point I was still going tubing quite often, to party and to climb, and I wanted to be tuber number one. I had been sixteen before but never number one. So I got up early, got drunk and then sliced my foot.

ROBSON: Probably because you were tuber number one?

MR SAKURA: Yeah, ha, ha, ha, ha, exactly. I was heading down to a mud volleyball court that was covered in mud and just by the river I stepped on a rock and slipped and it just… Pac-Man my foot… right through, not around, right through the meat.

That sucked and that was the end of that season. I just carried on partying. After I left here I spent a week in Malaysia. I was going to tour around but I just got myself a hotel room and let myself heal there for a week. The guy at the front counter was Bangladeshi and he just kept bringing me Bangla food. Malaysia is made out of tile. I had bamboo crutches. I couldn't go anywhere.

CHAPTER THIRTY-TWO

CALGARY STAMPEDE AND SWEET RIDE ANDY

MR SAKURA: So it started to piss down with rain, then I sliced my foot and then there's the Stampede.

I headed back to Canada in time for the Calgary Stampede so it must have been around June again. Probably back for another fucking wedding. People getting married a whole lot around this time of my life… dammit, I just want to hang out in Asia. July, I am in Canada again and I end up working at the Calgary Stampede, which is sweet. That was a real sweet money making opportunity.

I moved in with my dad who was living in Calgary at the time and I worked at the stampede as a doorman at one of the bars. People would be spending such a lot of money at that time and I was a doorman, just counting the number of people coming in and making sure there weren't too many. I'd always give myself space for about 20 more and stop letting people in until somebody would come up and say, 'Hey, here's fifty bucks, you wanna let me in?'

'Thank you sir, come on in…'

And then someone else would say, 'Here's 100 bucks let me and my two friends in.' I made thousands of dollars in ten days. It was as fucking stupid as that.

On the last day I really didn't give a fuck. I was going to make as much money as I could. I was still thinking in Lao money mindset. I was thinking, 'People are giving you a hundred dollars! Do you know how much a hundred dollars is in Lao? Fuck me, give me all your money!'

ROBSON: How much did you make?

MR SAKURA: Thousands of dollars in ten days. I think it was four and a half grand, and on my last day I made 900 dollars alone. I just had money falling out my pockets and I was wasted too. About a week later I received a letter from them saying, 'You are not eligible to work for us anymore.' I don't think they were too happy about my entrepreneurial spirit. I didn't give a fuck, I didn't care, I was going back to Lao.

For the rest of that summer I got a job landscaping in Calgary and met a German dude named Friedrich. He and I went for a road trip at the end of the summer. We drove all the way to Vegas from Calgary and then up the West Coast to LA, San Francisco and Seattle. That was cool; I got to see a lot of America in a sweet car. Sweet Car Andy.

We bought this car for 300 bucks and we got the mechanic from our landscaping company to fix whatever needed to be fixed which cost another 200 bucks. Then we spent 300 bucks on spray paint, made these stencils and then stencilled the fuck out of this car so every panel was different.

One would be zebra print, another would have flames coming off the tyres and another would have a flaming soccer ball. There was a big playboy bunny on the roof. The whole thing was just outrageous and on the front hood, 'Fuck it,' we thought. We both posed and stencilled our faces and so the front hood was black and silver with our silhouettes on. It even had my signature on it. It was just fucking pimp.

ROBSON: Have you got photos?

MR SAKURA: I will have to show you. There was a point I just stopped taking photos altogether and just started living.

We drove down to Vegas which is over 5,000 kilometres and it was a really good road trip. Everyone loves a road trip. I ended up showing up a couple weeks late for university. First day of class and I am in San Francisco. I call one of my buddies and ask, 'Hey, can you take some notes for me, I am gonna be a couple more weeks.' So yeah, went back to school late that year.

ROBSON: Here is the guy who, 'Spent half of last year in Lao and now he's showing up two weeks late,' yeah?

MR SAKURA: Hehe, yeah, and I was getting more and more focussed on getting back to Lao, back to Lao. I just wanted to get this schooling out the way and get back.

ROBSON: Why was it now so much more pressing?

MR SAKURA: I think because that was the second year I was doing it and we could see it getting so much better. By the end of that season there had been a couple more westerners getting jobs at bars and I thought it almost felt like I had to keep up my appearances. I have a part to play, I have to play this person. I have to act.

A lot of people were really stoked at how long I had been out there and just like were already stoked on my story of being there. Then they all left and I was lying in bed thinking, 'Maybe I should do that all the time. People are always going to be stoked on it, and it's only going to get better,' which it has the whole time.

Even when I was back in Canada, I would even be thinking about other flyers to do and different things about how to make the bar better. I was thinking more and more about it. You should see the doodles I have in my university books. It's all Sundowners Bar. I was always trying to think of different slogans.

Anyhow, September to December I am at uni and all I am thinking about is Lao.

CHAPTER THIRTY-THREE

MY BEST YEAR

MR SAKURA: Going back to Lao in the January actually turned out to be my most fun year. it was the best combination of doing all the climbing, tubing and partying. Red Bar opened up and Bar Uno which is now Life Bar was doing well too. The Red Bar team were awesome. That's where you get Jan and Sara… Jan and Sara, there you go. Jan and Sara, they were another couple to get married.

My buddy Indigo Zuma worked there too. He's a Canadian dude, a fucking legend, just super positive. He taught me how to do backflips off the swings and just… then there was Thompson, Anton Thompson. The guy behind Goosin' the Globe and the shotgunning prayer. That was all in Red Bar. Just a sick, sick team. They were really good promoters and really good at everything.

We really focussed on being the after bar, so once they closed we just focussed on getting everyone to come down to Sundowners. They were really taking off and they had the same vibe that Sakura has now.

It started off as a restaurant that just took off from there. Indigo, Thompson and Micky; these three guys showed up and just said, 'This is how we fucking party.' I think Micky is from Australia and Thompson is from Cornwall in the UK. They really came in and amped it up. They would all come down and hang out at Sundowners and they're all super. I was buddies with them all, the whole team in fact, and we would go on little road trips together. That's when I met Bob and Ben - that year was just amazing. It was the perfect combination. It was the beginning of the crazy party but it wasn't super crazy yet.

It was that time when we were building beer swords and wizard staffs, I don't know if we were shotgunning too much then. That came at the end of the year. I think it actually came with a dude named Shaun, a guy from Oregon. Then shotgunning really took off. They called this guy King Dreads and he was super amped on partying and everyone was really getting on board at that point.

At the end of that year I stuck around until after my Birthday for the first time. It just felt like my life was more here. That was mid-July. 26, my 26th Birthday. It was a mess. We opened Sundowners Two because of the flooding on the island at that time of the year. We had a good party going that summer and we couldn't just close because of the rain, so we built a bunch of huts over here on the mainland.

Fuck why did I go? Probably another wedding or something. I am trying to think what the song was. [Shouts, across the road] We're not open yet… people are just desperate to get into Sakura, that's the problem.

ROBSON: Looking for a T-shirt. Everyone wants a T-shirt.

MR SAKURA: Everyone does! July? Oh no, that's when I went to Europe, how old am I here?

ROBSON: 26.

MR SAKURA: That's right. When we closed up for the season I took Joey to Kuala Lumpur for his first holiday. It was his first time in an airplane, first time seeing a building bigger than five storeys. KL is quite modern and he had only been to Vientiane before. I then went from KL to London.

I met up with Disco Dave. Disco Dave, fuck me… he picked me up at the airport and he is from where the fuck is it? Oh shit… South London, tennis…

ROBSON: Wimbledon?

MR SAKURA: Yeah, Wimbledon, he's from Wimbledon, I went to

Wimbledon. I went there. and I went to Disco Dave's house and stayed with him. Dave Copthorne was his real name but when he was in VV he was Disco Dave. The man who never stopped dancing. He would dance in his sleep I swear, he was always moving. He was part of our team that season.

He's really good, really on it. Together with Hans, this German guy working at Red Bar. He was really technical and did all our menus, flyers and shit like that. He started at Red Bar and we ended up stealing him. He wasn't really outgoing enough for Red Bar because they were really getting on it. We took him in, 'Yeah, you're good shit Hans, come work with us.'

So after I saw Disco Dave I went to visit Hans in Germany. I flew into wherever he lives, a nice cheap flight on a low-cost airline and he lives right by the airport. That was sweet and then went to visit Jurgen just outside of Oettingen.

It was their 1000th year of being a village or some shit like that. I enjoyed hanging out in Germany and then I flew straight back to Asia, I didn't go back to uni this first semester so I came back to Lao in September. That was when they started flying low-cost flights from KL to London and it was stupid cheap to get a ticket. That made it easy to come back here.

I had to go back to Canada in the December to organise my university work. I sorted out my following semester to get course credit. I was home for Christmas and did a little stint at Two Dudes Pizza until I had enough to come back out in the February. It took a little longer to make the money because I had been away for so long.

ROBSON: Were you making money in Lao working in these bars?

MR SAKURA: Yeah, I guess, some. Fuck they were giving me a good percentage of the bar because it was getting busier, busier, busier, it was mostly me I guess, my magnetism if you want to call it that.

ROBSON: That was T.O.?

MR SAKURA: I think T.O. started around this time. We always used to draw on each other. This was around the time the magic markers came out and they hit the river real hard. Lots of writing, lots of funny slogans on people. We started getting a little more creative with the T-shirts when Hans started working with us. Hans came right back out too because he was loving working at the bar. He is back doing uni right now. I talked to him and told him he is missing out.

ROBSON: Is he coming back?

MR SAKURA: He says he really, really needs to.

RANDOM KOREAN MAN: [Walks past] Hey! Titanic!

MR SAKURA: Hey Soju! Titanic! Later! Later tonight! See you later, *annyeong!*

CHAPTER THIRTY-FOUR

INTO THE WILD WITH TEAM BUFFALO

MR SAKURA: I was back in Lao and this was another good year. I think this was when the shotgunning took off in a big way. It was kind of a continuation of everything that had gone on previously.

Beer pong. We brought beer pong here. I had a buddy from Canada come to visit. He said I needed some beer pong tables so he poured one out of concrete. It was a permanent fixture down at Sundowners.

This was also the year of our debaucherous trip to Cambodia. That was a good one. Every year when I came back there was always a little obstacle and this time they were razing the land on the island so we had to take down Sundowners Bar and rebuild it later. Apart from the beer pong table made out of concrete. That just stayed there whilst everything was taken down around it.

Whilst they were taking it down I had a crack team, Team Buffalo, go with me down to Cambodia. Team Buffalo consisted of Johnny Bravo, fucking legend. Another guy from London. He's six foot five and a super promoter. His name is Johnny Bravo… his real name is Dean. Dean Leonard The First. No, wait, Dean Holdstock. He's from North London, Tottenham. Super awesome dude. One of my favourite people I ever met in my entire life.

He was another guy who would always tell me, 'T.O. you gotta fucking smash it… you have to keep it going.'

I would be exhausted but would say, 'Ok, ok JB. I will.'

And I would… I am.

He was on that epic road trip to Cambodia in April, along with Hans and Bravo's girlfriend Toria. Twinx, this other chick from Brighton area and Rowan who was another Brit. Four Brits. Was it four? We got this guy to drive us down to Cambodia. We just got wasted on the whole trip down there. Everyone was naked in the mini bus. It was not very traditional Lao but we ended up in Cambodia so it was fine. We were just smashing the beer and really stepping it up.

ROBSON: Where did you go in Cambodia?

MR SAKURA: We went straight to Sihanoukville to party for a month. We did a fancy dress theme every day. It was something different, something new, every single day, usually every day and night. Every day we'd do a theme, every night we did a theme.

ROBSON: Examples of themes?

MR SAKURA: Pirates, Superheroes, Animals, Mix and Match kind of shit, Formal, any sort of theme that we could think of. We would get up in the morning, order two pitchers of beer and a little bit of food. We would send two people out to go out and buy drugs, two people to the market and two people to make stuff.

People who went to the market would pick up whatever random shit they could find. The two people staying behind would figure out whatever shit we were going to do that day. The two buying drugs, bought the drugs. We did that every day for three weeks in Sihanoukville. Disco Dave ended up meeting us there as well.

We came back through Thailand and it took us forever. I ended up back here. I must have been home at some point. I went home probably to write an essay. Then back to Lao in May. I have to go by my birthdays now because this one is super messy. This is when stuff was really kicking off. More shotguns, more debaucherous stuff.

ROBSON: That was when VV was really starting to take off?

MR SAKURA: It really was! The previous year a lot of really cool people showed up. That really got the name out so the following year was like everyone who went home, went and told their friends and their brothers and sisters.

'You should go to VV, it's cool go tubing.'

That year I think I might have spent my birthday in Vancouver, or was that my dad's because maybe that was the following year, anyway, something, I spent the summer back in Canada.

ROBSON: Wedding?

MR SAKURA: Yah, probably a wedding and it's nice to be in Canada for the summer and because I had been in Europe the previous year. I did first semester in Canada because that was one of the semesters when I had to do all the theory work. Whilst I had switched my major allowing me to do more independent studies, I did still have to do some theory work to get other credit.

ROBSON: At that point did it not seem like a waste of your time? Why do you keep going back to Canada? Your life is clearly in Lao.

MR SAKURA: At that point definitely.

But I was, 'Fuck, Stephen, just get this fucking thing done.' So I can live in Lao.

Every year back and forth, 'Ok, fuck, this is getting ridiculous, pick one!'

Still, I had to finish my degree. 'I've started, I need to finish, I am only a year away, only two semesters away, just fucking finish it.'

September, uni. Christmas at home and then January… back to Lao.

CHAPTER THIRTY-FIVE

THE DAY I RAN OUT OF MONEY

MR SAKURA: January and we're still at Sundowners. At this point Happy Bar is closed or not really competing at all and Red Bar is really hitting off. Space Bar was really big then too. It was on the island; it is now on the river, but back then it was on the island. There was a lot of competition because a lot of people were returning. That was a big one. We had to really step up our game then. We were doing more T-shirts, more advertising and more promotions than ever before. That was when the free buckets started. The bucket wars. We used to give out free whisky buckets, that was the start of it.

ROBSON: Did that fuck everyone up?

MR SAKURA: Yeah, yeah, but that was what heavy competition was doing to us. It was us and Space Bar also on the same island. We were the exact same kind of flavour, same layout, two bonfires, dance floor in the middle and a bunch of huts with hammocks. Their bridge was a little closer to town than ours so we were trying to divert them our way but then they had to go via Red Bar which was also open at the same time.

We gave out buckets 9-10 p.m. and then they give out buckets 10-11 p.m. so we would then do 9-10 p.m. and 11-11.30 p.m. It was just so ridiculous. I think in the long run it was actually good that everything got shut down in the end because this sort of thing has now stopped.

In total, we had three bucket wars. It got ridiculous. The people in charge were getting so obsessed with getting the most customers, they were not thinking how much money they were making, but that's quite typical here to not be thinking about business.

ROBSON: Were you getting fucked still?

MR SAKURA: Hell yeah, oh yeah.

ROBSON: Weed, mushrooms?

MR SAKURA: Not so much mushrooms at this point, and not so much climbing either hahahahah.

ROBSON: You were still T.O. at this point?

MR SAKURA: Yeah, T.O. is the rebel. 'Let's get fucked up. I am motherfucking T.O.'

ROBSON: You used to say that?

MR SAKURA: Yeah, as a persona. Every day I would go to the river, write T.O. with a magic marker.

ROBSON: Were you hands-on managing Sundowners or were you just a customer helping out?

MR SAKURA: It was more like the latter. Sakura is more proper management. Happy Bar, even when I became Daddy, that was more management, but T.O., I was just getting fucked up all the time. No responsibility. Threesomes. And yeah, really having a good one.

Bob and Ben came back and Ely, one of their buddies from home. I met a lot more really cool people and that is when it started to get really crazy. More tubing bars were popping up. More everything was popping up. The island was getting really crazy and became known as The Party Island.

That was the year when a lot of people died. That was the one when they quote the number 26 but it was the year after when the important person died. That was the big turning point.

ROBSON: What was a typical day like in that time?

MR SAKURA: Get up, have a bucket of… might have been, I don't think I was drinking vodka sodas yet… have a bucket, make a quick sandwich or something to eat. Head down to Sundowners, get dressed up, make some headbands and shit for the river. Get some spray paint, oh yeah, spray paint on the river was getting really creative. We made all sorts of stencils and shit on people's backs down the river. We would just spray paint the fuck out of everyone. Red Bar opened up a bar on the river. That was all that year.

We would be trying our hardest to get people to come down to Sundowners, promoting all along the river.

'Come down to Sundowners, free buckets from 9-10!' We were promoting, promoting, promoting. Then we would get back and usually have a bite to eat at Steakhouse. There was a point in time when we had to close Sundowners because we weren't allowed to play music on the island so I opened up a restaurant and I also managed Red Bar. That was nuts because those guys knew how to party.

I was sick of all the places here so I decided to start my own restaurant, a steakhouse and pizza place. I started renting it for a million kip a month which was very feasible and built a pizza oven. We got all that started and then as soon as we were busy, the landlady said the rent had gone up to 10 million kip. It was also tough because I had to get all my ingredients from Vientiane.

It doesn't sound like that big a deal when you first start but after you have been doing it a number of times, it sucks. It is just a really shit drive. It's not like just getting on a motorway for two hours, there are cows and dogs on the road. In Alberta, where I am from, there are just straight roads. I have no problem driving 200 kilometres at home. I would do that to get to the next city along just for a party.

The landlord's sister also started taking my ingredients for her own restaurant. She stole my menu too. She had all my menu items but didn't know how to make any of it, so once that happens, customers think you are all the same.

'I had that over here and it was shit.'

'Yes, but this time I am making it!'

A lot of the restaurants here suck at cooking 300 items on the menu. It's not a menu, it's a novel. How can anyone do all of those well? Pick 10 or 20 and do that. It was based loosely on what we did at Two Dudes pizza place. There we had 100 different pizzas; all the traditional ones but also some crazy ones. I took ten of their recipes and adapted them a little bit to what I could get here. I had some really cool pizzas. The steakhouse pizza had slices of steak, mashed potato, bacon and capers with a gravy base instead of tomato sauce, it was real… mmmmmm.

I had a taco kinda pizza too. I forget what I called it but it had salsa instead of tomato sauce. We had a disclaimer on the menu: Each pizza is hand tossed by a Canadian pizza wizard. Every menu item description was a little funny, a little tongue in cheek. 'The Burger Royale Deluxe Supreme – two patties so fresh… better than McDonald's.'

That is one thing I love about here; you can open up a business and you can fuck up and you aren't thirty thousand dollars in debt. I spent two thousand and we still had fun doing it. After Steakhouse closed we started playing music and slowly built up Sundowners again. Space Bar was doing the same. Eventually, we had the party back going til three o'clock or four o'clock and by the end of the season, we were right back to normal.

I spent Canada day, July 1st, here and then went to London for my 28th birthday - the big VV reunion party. All these people from the previous four years all met up in London for my birthday at the Hoxton Gallery – in a basement behind a green door. That was a really good connecting and reconnecting moment. We had a couple hundred people there with DJs and bands playing. It was really cool. Good times. People came from a lot of places to be there. Oh and that was when I found out Alannah and Craig are still together and Tim and Jenny are still together. Jan and Sara too.

I got to visit a lot of London folk there and spent my birthday just smashing it in London. I flew to Canada for my dad's birthday, his 60th. I wanted to be there for that and he was living on the East Coast by then so it was really easy for me to pop over from London. I spent two weeks in Canada, sobering up. It was a really, really heavy six months that one.

Then I went to Amsterdam for the first time and I loved it. I went to hang out with Bob of Bob and Ben fame. Bob van den Branden. He lives in a real fucking pimp house that doesn't even have an address; it is just called The Greenfields. He has a pool and I stayed in his pool house for a month.

I spent August in Amsterdam and then in September I went to Germany and hung out with Jurgen again and met up with Friedrich. He was the guy I went on the road trip to Vegas with, in Sweet Ride Andy.

He lives in Cologne so I stayed with him and did a clinical trial study in Maastricht which was real close by. I had to wear a cast on my leg for two weeks so they could see the effects. They took muscle samples and blood tests and so on to see what would happen with this cast. I got five hundred euro for effectively doing nothing.

I guess that was the beginning of October. Then, I threw a Halloween party pub crawl in Amsterdam and I think that is where all my money went. Since I had had such a good time in London for my birthday, I wanted to get as many people as I could together again and keep all the connections going. It wasn't quite as big as the London event but we still had forty people show up from all over. Lots from Germany and London, as well as Asia and Canada. I sublet an apartment off one of Bob's buddies, a one bedroom flat, and all these people crashed there. There were a lot of people sleeping on the floor and a lot of people not even sleeping, just getting fucked up.

Since there were a lot of people from London, a few convinced me I should come back. My flight back was in December from Paris to KL but I had time. I had wanted to work for a pub crawl in Europe or something but it's tough to do that. Who would have thought?

'Yeah I will just fucking work in Europe and yeah I will make money and it will be fine.' But it didn't really work out like that so I went to London and met up with Johnny Bravo. He got me a couple promo gigs, flyering at Wembley Stadium. Fifty pounds for a couple hours work.

The problem was the money from the clinical trial hadn't gone into my account yet because I had one final check-up the day before I was flying. I was dependent on getting this five hundred euro to sustain me.

From Maastricht, I needed to get to Paris to get my flight home. I had booked a train to Brussels and a mega bus, or whatever the fuck, from there to Paris. The train from Maastricht was cancelled so I ended up missing the bus. The cheapest train ticket was 110 euro and I only had about forty or fifty on me. I tried to explain that my train had been cancelled but the woman at the ticket counter was unsympathetic. She said because it was a train from a different country it had nothing to do with her. I even asked if I could use my mum's credit card over the phone but she wouldn't let me do that either.

'I am going to have to beg for this money.'

'Ok, fine. You have forty-seven minutes.'

'Fuck off.'

I went and asked people for money. This is the first and only time in my life I have ever begged for money and I made just enough, just in time. High five. I got to Paris on this 100 euro fucking train and I had fifteen euro in my pocket at that point. I get to the fucking North Paris train station and still needed to get to the low-cost airport. I took a bus, got off at the wrong stop and ended up walking the rest of the way. It was the beginning of December and it was fucking freezing. I had no idea where I was, I could just see the occasional airplane going over.

I made it just in time, but I only had two euros in my pocket. Fuck I was hungry. I realised I was going to land in KL and still only have two euros in my pocket. December 4th. The day I ran out of money. That day sucked. I still needed to make it until the money arrived in my account from the clinical study, or my mom to show up and that wasn't going to be for another ten days on the 16th. Fuck, fuck, fuck, fuck, fuck, fuck, fuck, fuck.

I am sitting on the plane and I notice these three guys sat next to me, dressed real funny in a traditional getup. These guys must have a story, otherwise, why would they be dressed like this?

'Hey, what's your story?'

'We are tradesmen from Germany. It is like an apprenticeship - we are on this journey where we aren't able to work for money, just for accommodation and food. This is our thing. We also are not allowed to return to our hometown for two years.'

'Oh really? Guess what? I have a bar in Lao and I need a tonne of stuff to be built. You can come up there and I can get you hooked up no problem.'

'Sounds great.'

'The only catch is: I have no money right now, like zero. Do you mind hooking me up for the next two weeks? I can show you around KL because I have been there a bunch of times before. If you guys hook me up, I will pay you back as soon as I get the money, plus as soon as you get to Lao, like, find my bar and you have as much work as you need to do.'

'Sure.'

'Fuck yeah.'

YES! Thank fuck I bumped into them. I hang out with them until my mom and her boyfriend show up. This was going to be my first Christmas away from Canada but since my mom was there it was fine. We hung out in KL, spent Christmas in Langkawi and then we all flew together to Lao.

PART FIVE

DADDY

CHAPTER THIRTY-SIX

GO TO JAIL, DIRECTLY TO JAIL; DO NOT PASS GO, DO NOT COLLECT 200 KIP

MR SAKURA: It was a very nice time over the festive period with my mum and her boyfriend. We spend new year's here in Lao. Joey had started running Happy Bar by then so he had split from his brother, who still managed Sundowners. We got back expecting to work for Sundowners but when Hans and Chula arrived in Vientiane, Joey explained to them what had happened. Chula was a big supporter of Sundowners as he had been working there for the previous six months. It was weird to suddenly be working for Happy but it had been closed in the recent past leading up to that and it was the same people running it.

We had the slogan 'Drink Triple, See Double, Act Single' on the T-shirts for Sundowners Bar. We had just come up with it as we were switching bars. I think there are probably like six shirts in existence that have 'Sundowners Bar - Drink Triple, See Double, Act Single.'

ROBSON: Wow, where are they?

MR SAKURA: Chula's got one, Hans has got one, Joey's probably got one someplace.

ROBSON: Did any customers get one?

MR SAKURA: I don't believe so. There might be a few out there.

ROBSON: That's even more rare than the white on pink?

MR SAKURA: Oh yeah, that's like, if you can find one with Sundowners Bar and 'Drink Triple…' you'd be doing well. Actually, I think Dean might have one.

ROBSON: I want one.

MR SAKURA: Good luck!

ROBSON: Presumably Happy Bar T-shirts are pretty rare too?

MR SAKURA: Rare-ish we had a couple of funny ones, like pirate theme days with Happy Barrrrr with a bunch of Rs and the face of a pirate. These are the rare ones. You'll see lots of locals have them. When we all went to jail we had a big bag of T-shirts which got 'distributed'. I had no idea where they went but obviously someone got a hold of them. I'd be walking down the street and see a local wearing one and thinking, 'You probably didn't drink two vodkas…'

At New Year's I was at Sundowners but soon after I was all about Happy Bar. Joey went to Happy, his brother stuck with Sundowners. I hated Happy because it had always been the enemy. Like going from Real Madrid to Barcelona. But Joey was my brother. Well, they were both like my brother but I was always closer to Joey.

ROBSON: Do you remember the day you changed from Sundowners to Happy?

MR SAKURA: It was a gradual thing because Chula, Hans and I tried to do both but it was weird competing with ourselves. We were trying to get everyone in Sundowners and then everyone in Happy. It didn't really make sense. We ended up confusing people and we were still trying to compete with Space Bar. It was crazy to compete with ourselves when there was more competition on the island. Every time we had people from Sundowners from previous years, they would tell us it just feels right to stick with Sundowners. It was a grungier bar and T.O. was grungy. Happy was a bigger bar which was better done and put together nicely. By February, it was just Happy which is where the St Paddy's Day T-shirt came from in March about three years ago.

ROBSON: You were only in Happy four months before you got busted? Ironically if you stayed at Sundowners you might not have got busted?

MR SAKURA: I think it was going to happen either way. One of us, Space or Red Bar were going to get busted first. At that time of year, we were just doing better - we were winning. After all that work with the T-shirts and the buckets and the flyering, we were the busiest. People were saying they had seen our shirts all over. We were doing it, doing, it doing it.

ROBSON: One minute you are owning it and the next you've got the cuffs on?

MR SAKURA: Exactly. It was a really good year up until then. Things were really picking up and I remember when we did our first 10m kip night and we celebrated the next day.

It was Chula's birthday and we rented these ATVs. They were brand new; a guy here just ordered them to start renting them out. We took them straight off the truck. Fuck, we just bagged the hell out of them, you know when you get a new vehicle, 'Let's see what this fucking thing can do?'

It was the rainy season; lots of big puddles, mud everywhere, we had a great day, it was fantastic. Just tearing it up.

ROBSON: When was that?

MR SAKURA: About two or three weeks before we were shut down.

ROBSON: So you were on a high?

MR SAKURA: Oh yeah.

ROBSON: You were thinking the adventure tourism is going to start here?

MR SAKURA: Yeah. Everything. We were going to grow VV big time. The bar was going well, we'd moved to Happy which was easier to get going. Everything was really looking up. I had a slight inkling the cops were going to be around, but just local cops.

They would usually pop in and say, 'Hey, just make sure no one is smoking in the huts.' I thought I was safe in the kitchen.

ROBSON: So Joey was cool with you selling other stuff on his property?

MR SAKURA: We would always ask the question, 'Hey dude can we do this?'

He'd be like, 'Yeah, I think so…'

ROBSON: You knew it was coming?

MR SAKURA: It was an eventuality for sure. Doing that sort of stuff, once it gets too big, will land you in trouble.

ROBSON: Especially when the likes of The Guardian newspaper in the UK are writing articles about it.

MR SAKURA: Exactly. I knew it was going to happen, I just wasn't expecting it that night.

CHAPTER THIRTY-SEVEN

THE WORLD COMES CRASHING DOWN

MR SAKURA: I used to get in trouble every year; they would hit me up at least once every season.

ROBSON: They'd just rock up and say give me some money?

MR SAKURA: Well they did that regularly anyway but usually I'd be drunk or smoking a joint when I probably shouldn't be smoking a joint. After we close the bar, smoking an opium cigarette.

They'd be like, 'Stephennnnnnnnn…'

And I'd be like, 'Hey guys, what's happening?'

'What's that?'

'Ahhh fuck.'

'Ok, let's go to your guesthouse and get some money.'

ROBSON: Is that what they said?

MR SAKURA: Yeah, it's either my passport or some money, right? Hey, do you want a smoke?

ROBSON: Is that an opium cigarette?

MR SAKURA: No, [looks at it perplexed, considering it] I don't think so. No, no, no, I tried to be real good this time round.

ROBSON: Not doing drugs?

MR SAKURA: No, but I used to be a proper Canadian and always have a joint going. It was pretty easy to bust me smoking a joint, all you had to do was come down to the bar.

ROBSON: Oh I see, so if they were ever short of cash, 'Where's Stephen? Ah yeah, he's smoking.' How much would you have to pay?

MR SAKURA: It varied between weed and opium. One time they went after me because I didn't have a business visa, which wasn't drug related. Another time we got busted for doing some promoting using a big speaker on the back of a truck.

'I guess you're not allowed to do that? No one said we couldn't.'

There was a big International conference being held later that year and that was when everyone was expecting the crackdown to come, not in July. The crackdown came because of the famous son. I think that sped things up. That happened in April. Considering how long it takes for political stuff to go through, two months is pretty quick when you think about it.

'We are going to cut you off or you need to do something about Vang Vieng,' so I think that is what sped it up. Admittedly they were probably going to do it anyway for this conference. So July of that year, everything came to a head and my world came tumbling down all around me.

Game over. Insert coin.

ROBSON: Another Pac-Man moment.

MR SAKURA: The government came up here. Everyone got in a lot of trouble. They shut down all the bars and everyone went to jail in one fell swoop. No one knew it was coming.

It was a regular night and it was even starting to rain. Usually, the police don't come out in the rain so I should have thought something was up. I was hanging out in the kitchen at Happy Bar. There was a plate of mushrooms next to me and a bong when these Lao guys come in, not even dressed up and started taking pictures.

'Woah, what are you up to?'

'Police, police, police, police!'

I got a real sinking feeling in the pit of my stomach. I didn't recognise any of them because they weren't our usual police. It turned out they were from Vientiane. It was at that moment I knew the game was up. I threw the bong out of the window, not that that made a difference and I was like, 'Oh fuck.' That was how that happened. At the time we were in the middle of a good thing and that halted pretty snappily. Everyone from Happy Bar went to jail.

Jail was your average third world kind of prison. I was the only foreigner in there together with Chula, the slogan guy who was Canadian but half Asian. The rest were from Lao. Fourteen of us in one cell about the size of a regular guesthouse room. No bedding, just a concrete floor. At night or when the guards weren't there to watch us, everyone had to put one foot in a stock. Imagine a big piece of wood in the centre of the room with round notches in it. Everyone places their ankle in a notch and another piece of wood is placed over the top and it gets locked down.

Seven of us, on each side of this big piece of wood, sat across from each other. Each person lying on their back with one leg locked into the stock. I couldn't really move around much. If I needed the toilet I had to piss in a cup or water bottle and I had to shit in a bag. One guy was passing a kidney stone. Every time he'd go to piss he would shout out in pain, he really was not having a good time.

There was a squat toilet in the back corner and a little wall about three feet high to give you 'privacy' if that is what you could call it. I mean if you thought the toilet at the bus station was bad you should have seen this. We used to get half an hour of running water a day. That was just about enough time so that everyone can have a quick shower and fill up the water bottles and flush the toilet. As you can imagine, it didn't exactly smell of perfume and meadows in there.

For food, they gave us two packs of instant noodles and two little cartons of soy milk to share between us. We would stand in a line and one person would break off a small handful of noodles, crush it up, add some salt and wash it down with the soy milk. Then he'd pass it on to the next guy. That's mama's noodles for you. We only had one meal a day unless you knew someone on the outside who could bring in stuff. We were lucky because we had Joey's family. They brought us tonnes of sandwiches but they would never get to us. It starts out as one big bag of sandwiches and everyone along the way gets some of it. By the time it got to me I got one sandwich and even then had to share it.

I was sitting in jail and no one had come to talk to me for three days. I remember on day three thinking, 'Oh, fuck tomorrow's my birthday and it's Friday. I really hope something happens by then.' Everyone was telling me if we don't get out on Friday we would definitely be staying in until Monday.

CHAPTER THIRTY-EIGHT

MORE THAN THE GIFT OF FREEDOM ON MA BIRTH-DAY

MR SAKURA: 'Get out of town and cut your hair.'

ROBSON: Is that what he said to you?

MR SAKURA: Yeah, 'Look different and do different next time.'

The main policeman told me that when I gave him all my money and he let me out of jail. It cost me about 8 million and my iPhone. He just looked at me and said, 'Is this a 4? What? Not a 4S?'

On the plus side, they let me out on my birthday. It was sweet. Friday 13 July. Best birthday present ever. They let me and Chula out of jail on the same day. Poor Chula. He got arrested three times that year and it was mostly because of me.

Before I had got myself arrested, I had been seeing this one girl from Lithuania. We had been hanging out together although we hadn't slept together. The day they let me out of jail, my birthday, was the day she was supposed to be leaving. I had spent the last four days thinking about her and was just hoping I could see her again before she left. Everything had closed down so there wasn't really any reason for her to stay. She might even have gone already. I basically left the prison and just ran to the bus station. The bus was already started up and she was loading her backpack onto it when I managed to reach her. Luckily buses in Lao never run to time. Have you noticed that?

ROBSON: Once or twice.

MR SAKURA: 'You cannot leave today! No way! I just got out of jail, it's my birthday, we're going to fuck like rabbits!'

She looked a little shocked to see me.

'I've been thinking about fucking you for the last four days in jail, like, just like, please.'

'But I am getting on this bus…'

'No you're not!'

'What?'

'I'll buy your next ticket out of here whenever… we are fucking tonight!'

She thought about it, shrugged, and said, 'Ok, sounds good to me.'

I was like, 'Really?'

'Yes.'

'It's ma birth-day!!!! It's ma mother-fucking birth-day!' I danced like a crazy motherfucker all the way up the street, 'It's ma Birth-day!'

That was cool. We unloaded her bag and went out partying. There was one bar left open - Space Bar, which had survived the clampdown but they still had to close at 11.30. When they closed we were walking back and stopped at the Sundowners Bar/Happy Bar bridge. Both of the bars were closed and there were no lights on and no one was there.

'Hey - you want to go fuck on that bridge?'

'Sure, let's go for it,' she winked at me.

That always really turns me on, the exhibitionism thing, and she was up for doing it wherever. We headed down there and we start going at it on the bridge and she's bent over the railing looking down the river. I am thinking, 'This is fucking awesome.' A whole bunch of tourists walk past and I'm like, 'Ooooh sorry…' Then one of the cops that let me out that day comes walking up one side of the bridge.

'STEPHENNNNNNNNNNN!!!'

'Oh fuck!'

He was so pissed off.

'You wait, you wait! My boss come, my boss come now.'

I had to wait for the police chief that sat me down in his office earlier that day telling me to get out of town. This was a whole lot worse than smoking weed… oops. His boss comes down.

'Stephen, what are you doing?'

I still had my pants round my ankles and was trying to look normal. Pulling them up and tucking it in a little bit. He is also incredibly pissed off and I had already given him all of my money. He even had my phone so there was nothing I could give him.

'Uh, sorry… it's ma birth-day?'

'Ah, it's your birthday? Are you leaving tomorrow?'

'Yeah, yeah, yeah… yes! I promise. I am leaving town tomorrow, sorry.'

'Ahhhh, you go home now.'

'I'm so sorry… wait, really?'

'Go home now!'

I wasn't going to wait for him to tell me a third time. I take the girl by the hand and take off home and continue from there. That was a scary one. I thought I was going straight back to jail. I would have been fucked because I had no money either, it was a real close call that one.

ROBSON: Was he laughing about it? Did he see the funny side?

MR SAKURA: I think he kinda understood, he's not a horrible guy, but yeah, it is kind of offensive, and I know that. I had been here long enough to know and especially on the day I had been let out of jail, of all the days. Maybe that wasn't one of my smartest moves but it seemed like a good idea at the time.

ROBSON: Was she scared?

MR SAKURA: It wasn't really about her, it was about me, I was the wanted man… I think she kind of thought it was kinky; a wanted man running from the law. Girls love that shit, getting busted by the cops, getting out of jail, a rogue, leaving town tomorrow, make it count.

ROBSON: Did you leave town the next day?

MR SAKURA: Oh yeah. I went to Luang Prabang the next day. First bus out of town.

CHAPTER THIRTY-NINE

GETTING MY ASS KICKED BY LIFE

MR SAKURA: The whole Happy team went to jail. We had all been in the same cell. We lived as a team and we died as a team. So they let Chula and me out on my birthday which was great but it was up to us to bail everyone else out. 12 of us in total. That's when I moved to Vientiane as a teacher and while I was there pretty much sending messages on social media to everyone who had ever been to Sundowners or Happy, or who worked for us and who had said, 'Oh hey if you ever need anything just let me know.'

Now was the time I needed something so I would message them and say, 'Hey guess what? Everyone is in jail. Can you please help us out and send some money?'

I spent a year doing that. It really beat me down. Especially when people are like, 'Yeah, I'll send you some when I get my next paycheque.'

Then a few weeks go by and I am sending another message saying, 'I hate to ask again but everyone is still in jail.'

I was spending all my own money getting people out of jail as well. It took months and months to raise the money.

It was a mission and it was really disheartening so after a year in Vientiane, I was like, 'Oh fuck. I am broken.' I had managed to get everyone out after a year except Joey.

ROBSON: Did anyone from any of the other bars go to jail?

MR SAKURA: Sundowners wasn't open that night so no one went to jail from there. Happy was the busiest bar at the time and so we were the first one to get busted. They raided us first and everyone else kind of got given a warning. We were made the scapegoat. We were the busy bar at night, they knew people did alternative stuff there. The reason everything got closed down was because of the deaths on the river which really were nothing to do with us. We were open at night and no one ever died at Sundowners or Happy but we were made an example of and got shut down first.

Every few weeks I would come up from Vientiane with some money to get another person out. It was so depressing. The town basically shut down because there were no bars and therefore no tourists.

ROBSON: Why didn't you go to Canada at that point?

MR SAKURA: I had to get everybody out of jail.

ROBSON: Would it have been easier to do that from Canada?

MR SAKURA: Not really, I had no money and I didn't have a plane ticket. It was quite the predicament. I wasn't allowed to teach in Canada either so I would have done some construction job. I probably would have ended up dealing drugs in Canada and I had learned my lesson.

It's the only way you can make money, like the only way you can make real money, which is why this is so shocking and awesome. I tell my mom on the phone, 'Everything is so good. It's perfect, this is great and we aren't even selling drugs.'

She'd say, Really?!'

'Yeah isn't that crazy?'

'That's fantastic!'

After a year trying to get Joey out, I was just exhausted. Everyone else was out except Joey. He was the boss so he was going to be there for a while. I had to go back to Canada in the end because the whole thing really just kicked my ass. I was just really frustrated and annoyed with everything. I had spent six years helping other businesses and I thought I was part of the community. That seemed to count for nothing and I was just getting a real harsh time of it. That was a real kick in the pants.

I was teaching Lao adults English in Vientiane. It was dire, soul destroying. That year, that was a rough one. Coming off the back of my best year. Awesome year, awesome year, awesome year, even better… then kaboom… 'Fuck you!'

'Ahh, thanks for the help, take the fuck off!'

That's when they told me to leave for a couple of years, cut my hair, thanks for your phone, 'Why isn't it a 4S? We will just keep your bag of T-shirts.'

I did that for almost a year. A whole year in Vientiane was quite humbling. I was broke and always asking people for money. I hated having to do that and I was doing it daily.

ROBSON: You might have been quite lost in that year then?

MR SAKURA: Oh yeah, I lost my business, I lost my job. All my friends were in jail, the other workers had left. Tubing was finished, all the bars were closed, the community was gone.

When I'd come up here none of the businesses were making any money. Everyone kept asking me when are the tourists going to come back? It was really depressing. That is the reason why I want Vang Vieng to grow up and have a bit more depth to it. When it is shallow, all it takes is something like that and then it's all over.

CHAPTER FORTY

LOST, HUMBLED AND BACK IN CANADA

MR SAKURA: I left Vientiane just before I turned 30, which was a year to the day after they set me free. I wanted to be back in Canada for my thirtieth birthday so I went back there.

ROBSON: Did you feel after the teaching in Vientiane that you needed to go home?

MR SAKURA: Yeah, time to grow up, get a job, settle down.

ROBSON: Why did you want to do that in Canada?

MR SAKURA: I just wanted to have another party and have something to pick me up a bit and forget about Vang Vieng.

I was thinking, 'What the fuck? It's my thirtieth, I want to do something for it.' I literally had one whole horrible year. In fact, my entire 29th year was shit. 28th was the best, 29th was the worst. 30 was rebuilding. 31 was pretty good; it got me back here.

I had my thirtieth birthday in Vancouver with a lot of my friends from here. Everyone I know in Vancouver I have met here because I am from 2,000 kilometres away.

I decided I was going to pursue being a teacher. There were elements of the teaching in Vientiane that I enjoyed and I had my degree by then. Oh yeah, at some point around that time I finally got my degree. I kept telling people I've been going to uni for six years but when I thought about it I had been in Lao for five and l was still in my fucking twenties.

ROBSON: Did you even go to the graduation ceremony?

MR SAKURA: No. I would have been too busy enjoying life in VV to go back for it. I did get my degree though. I felt I needed to. Do it for mom. You know? Do it for friends. Show everyone I can actually get a degree. May as well have a degree than three-quarters of a degree. That way I am not the drop-out stoner who runs a bar in Lao - I have a fucking degree so eat it. That was one expensive piece of paper, Jesus. Most people get a BA and do nothing with it or get a desk job. The thing with me going back to Canada at that time was my degree was in fucking anthro and religious studies.

ROBSON: What did you plan on doing with that?

MR SAKURA: I didn't know, perhaps the same thing I was gonna do with a Psych degree. Get a job in Alberta where there is lots of oil and gas and stuff like that. I would have got a job working for some oil company doing something, who knows, just a nothing desk job. Fuck doing that so then I was thinking about teaching and going back to university to get a teaching qualification.

ROBSON: Talk to me about your mindset back then. You're back in Canada. Are you thinking this is where I am going to live? The dream is over; Vang Vieng is dead.

Are you thinking, 'Do I find another Asian backwater that is going to be on the next tourist trail?' or are you thinking, 'Just get my teaching qualification, be a Canadian, shave…?'

MR SAKURA: I was thinking very much, 'Get your shit straight, grow up, it's time. You had lots of fun. The Vang Vieng I knew and loved was over, so I may as well just die with it.'

ROBSON: The two years in Vientiane and when you were back in Canada were really rebuilding years. You had two really down years, out of the mixer, out of the party zone. Was that when you really grew up?

MR SAKURA: Yeah, I think so. I was real broken. Those two years. It was tough living as a student again after not being there for three or four years. It was hard being ten years older than the other students. The whole thing was just weird. I didn't want to go back to Two Dudes pizza, where I worked when I was 21. The whole place, that whole time was just grinding me down.

Then in the September, it turns out I am too late to apply for the university I wanted to go to, so I got a job at the university as a tutor. It got me back in learning mode at least, going to classes and helping students with learning difficulties. I am working in the university, back in my hometown, and everyone is married and has kids and is generally just boring as fuck. I hung out a little with a couple of other 30-year-olds who I never really got along with because they were in a different mindset altogether. It was straight up awkward. I was just working, drinking a bunch, doing a lot of cocaine and trying to keep up… On top of that, it was the worst fucking winter ever, even worse than normal, really shitty. The only saving grace was that I was part of a bowling league with a bunch of old guys. It was fun; we just got wasted and went bowling.

ROBSON: Did you go climbing?

MR SAKURA: A little but not much, I probably hadn't picked up my equipment in about 5 years before that. A couple of times I tried but it was like, 'Fuck, I used to be able to do this no problem and now I have jelly arms and a beer belly.' I was quite out of shape because it was the first time I had been in Canada for years and I was eating all the fast food I had missed out on. I put on quite a bit of weight.

In January I was finally able to enrol as a student teacher. I taught little kids twice a week and then the other days I was learning how to be a teacher. A mix of learning by doing and theory, it was cool and at that point, I was quite keen on it. I was thinking, 'Do this one and then it's only three semesters until you can be a teacher the whole time.'

I had two great kids and I was told I could talk about whatever I liked so

I ended up telling them all about Lao, Southeast Asia and travel in general. I was trying to get these kids interested in thinking about the bigger picture, the wider world and stuff like that. Around that time I was only just thinking about Lao anyway.

I would always talk about myself in the past. The number of times I would start a sentence, 'I used to run a bar,' or, 'In Lao, I used to… I used to…' I fucking hated it. I had been part of the community in Lao for years and I didn't want that to just be a phase. I just kept thinking to myself, 'Fuck there's no reason other than family for me to be here, I should go back home, back to Lao, and stay in Lao. Lao is my home.'

ROBSON: Your heart is in Lao?

MR SAKURA: Yeah, yeah.

ROBSON: It's almost like you are an honorary Laotian, do you feel Laotian?

MR SAKURA: I wouldn't want to say it but others may say it of me. Joey might say, 'If you're Lao… you need to learn a bit more language.' I know most of the conversations I need to have but I never took any classes in Lao or Thai, just picking up what I need to know to do my job here. I certainly feel more at home here than I did in Alberta when I was learning to teach.

After the first semester I got through to May and it was still snowing. It had started snowing in October and continued until June 1st. I couldn't believe it. I was thinking, 'I can't do this. I am not doing another fucking winter in Canada… ever. This is fucked up. This is not cool.'

The thing was no one else understood because it's always like that. They would say to me, 'Where have you been bro? This is winter in Canada.'

It was so frustrating, 'How do you deal with this shit? It snows all the time. That's why I haven't been here for fucking ever!'

That's when I had enough. I thought, 'Nah, I am going to go save some money and then go back to Lao.'

June 1st, I moved to Vancouver Island to get away from the snow and hang out with Chula. He told me there was lots of work there so I could make some money landscaping. I basically spent the whole summer digging trenches in the rain. I went to a couple of festivals, which were kinda neat. I tried to make a bit of money selling weed but everyone sells weed there so it didn't really end up working out.

I left Chula and moved to Vancouver, moved in with my buddy Stew and started working three jobs. I was teaching English to Korean children in the day, then worked for a food delivery service at night time, whilst also doing construction for a set design company.

Joey had gotten out of jail after 18 months inside. At some point, whilst I was in Vancouver he rang me to find out when I was coming back. He said it was starting up again. It was only smalltime, a few people just chancing their arm and Phat Gorilla had become the main bar.

Jane, now she was another character, a whole other story. She got married to a Lao guy and opened up Phat Gorilla. They luckily weren't working that night we got busted. I don't know if she really liked me or I her, but whatever it is we don't talk. She still calls me Boss Man which I can handle. In the interim, they are running it and Joey was looking around for something to do because he always had bars on the island and now that wasn't happening. He was popping around, helping Phat Gorilla I think, but there can't have been much action there, just the seedlings. It was in October last year that he got Sakura.

ROBSON: When did the tubing bars reopen?

MR SAKURA: They only closed for a couple months but none of the bars were meant to be open. New regulations came in a summer ago. They mandated closing times and said only five at a time could be open on any one given day. They had meetings to figure out how they could revive tourism in VV without fucking it up.

ROBSON: The government wants tourism?

MR SAKURA: Oh yeah, they just don't want people to die.

ROBSON: That mindset, you were working these three jobs, you're in Canada, 'Fuck this shit, my hometown is dull as, it snows, I am landscaping, I hate it, I am learning to teach English, I haven't got my qualification blah, blah, blah…' At what moment in December did you go, 'Fuck all of this shit I am going to go back to Lao?' As you said it was starting to build up and trickle, people were asking you when you were coming back and you had missed a season.

MR SAKURA: Yeah, pretty much the whole summer I was in Vancouver. All the people I was hanging around with then were people I knew from here. Pretty much the whole time we were talking about Double V and Lao and this one guy, who was selling me all my weed, he was like, 'We have to go back!'

He just wanted to come back for a month or six weeks. I was like, 'Oh yeah, I'll go back, I'll save up and we will go together. Let's go for six weeks.'

PART SIX

MR SAKURA

CHAPTER FORTY-ONE

RETURN OF THE KING
ARRIVAL OF MR SAKURA

MR SAKURA: Coming back I wasn't really too sure what the response was going to be. Remember, when I left the police chief had ordered me to leave, get out of town for a couple of years and cut my hair. I had no idea really what to expect.

ROBSON: Was Joey mad at you because he went to jail?

MR SAKURA: That was something I wasn't too sure about either. The last time I had seen him we were in jail together. I left before he got out because I was just sick of working in Vientiane.

ROBSON: Did you pay the money that got him out in the end?

MR SAKURA: I believe so. I think he just had to wait it out a little bit.

ROBSON: You had stopped dedicating your life to getting him out?

MR SAKURA: Yeah, exactly.

ROBSON: So you didn't pay the final bit that got him out.

MR SAKURA: I'm not sure.

ROBSON: But there wasn't a beautiful ceremonial moment when he was released and you had a big man-hug?

MR SAKURA: No, not at all. I had left by the time he got back. But what else was I supposed to do? So coming back, I wasn't sure what to expect. A lot was reliant on how me coming back was. It didn't really matter if I was going to be a part of Sakura or not, I just wanted to be back here, just back around VV or in Lao, that was all it was.

When I got here, though, it was awesome. Joey and I went for a walk and everyone was super pumped. 'Stephen's back!' it was sweet. Cool.

'I am gonna stay here now.' That was the big one. When I came back in January-February of this year I actually felt part of the community, it was nice, something I didn't get in Canada.

I mean maybe I had friends and stuff but even there everyone is just like, 'Stephen, what are you doing? Are you living in Asia? Are you gonna stop doing that at some point?'

I was just like, 'Whatever, I don't know.' I really wasn't too sure until February. To cap it off, the success of Sakura going so well is a nice little bonus, like sweet. It's a tonne of fun. I don't plan on leaving any time soon.

ROBSON: You don't?

MR SAKURA: No.

ROBSON: Are you going to die in Lao?

MR SAKURA: Well, I hope not. That's the one thing, I always get a little worried when I get a bit sick, but I have some doctor friends at home. I send them my x-rays and blood test numbers. Their diagnosis is usually that I drink too much.

ROBSON: So Joey had taken over Sakura in the October and you returned in the following January?

MR SAKURA: Yeah, Sakura was nothing at that point. It wasn't even a place to watch movies and eat. It was for rent. Joey had started renting it. He was like, 'Hey, why don't you come back, there's space on the team.'

I was like, 'Fuck yeah. Sure yeah, let's do it this time. But we're not doing any alternative stuff.'

I came back to work and it was starting to get busy. It was just like the old days. I showed up on a similar sort of day, January 12th I believe it was. It was the exact same business model; free T-shirt. We weren't allowed to sell buckets anymore because there were a lot of regulations due to the crackdown but it actually worked out a lot better. In the meantime, there's this Korean TV show that showcased VV and all of a sudden all these Koreans were coming and they love to party so it was sweet.

They'd already been going a couple of months but since then we have really just stepped it up and been smashing it since February. The feeling we have is, 'Let's like, really do this.' Sakura is in its infancy. You should have seen the floor when I got here. It was in a state, everything was falling apart. No one really knew what steps to take. I knew some things that had to be done and then there were these German carpenters walking down the street and I know exactly what they were up to. Traditional journeymen. Traditional coveralls and black hat. The same type of guys working for me in Happy Bar and built some really cool structures.

These are the guys that travel the world and aren't allowed within a 15 km radius of their hometown for three years and three days. They're out to learn from other masters from different areas and different ways of doing their trade. Before, it was maybe a couple carpenters and a stonemason, this time it was a couple carpenters. As soon as I saw them, I was like, 'How long are you guys here for?'

'We are not sure.'

'I got your food covered, ok, I got all your food. I got your drinks too. Anything else? Your accommodation? Covered. We need you to do everything. Like, EVERYTHING!'

Back then, when we played Gang Nam Style the place would fall apart. Someone would jump on a floor board and it would tip the fridge over on the other side of the bar. They built a brand new bar in the back, took out a few walls, replaced the floor. They really helped out there. They rebuilt this bar with solid German engineering.

The other guy who was management at the time, he was like, 'I don't trust these Germans.'

I was like, 'No man, just trust me, trust the Germans, let them do it.' They just made the bar. They went into the jungle, about ten kilometres from here and got these pieces of wood. They carried them through a cave. My buddy rode them down the river and then they treated it. I had loads of stuff for them to do. They had to leave in March but they did a lot in the time they were here.

So the whole summer I was planning on coming back, I wasn't sure if it was going to be for six weeks or for good. It was all going to depend on how my arrival went down with the cops, Joey, the community and just about everybody. I was really worried it was going to be weird coming back here and it was going to be awkward like it had been in Canada. So I had a return ticket for February 28 or something like that and I was sitting at the Irish Bar when that fucking plane took off. I was supposed to be on that flight but instead, I was eating breakfast. At that time I had made my decision, 'I am not going home, fuck that.'

And Sakura was picking up by then. When I came, there was a manager here named James. He was here for a bit. I was shadowing him, seeing what he was up to but I wasn't officially the manager at that time. I was Daddy though, and everyone working knew that. It was the exact same team as Happy Bar, so they all sort of knew who the manager really was. The ones I didn't know were confused as to what was going on.

My response would be, 'Well, James hired you, listen to him, but if you ever need anything, done talk to me. Then after a month, even though I never said it, everyone knew who was actually in charge.

ROBSON: Was there a power struggle? Did he leave because of that?

MR SAKURA: No, no, I told him what my position had been and he said, 'That is my job now.'

I said, 'That's ok, you can have it. Well, you're leaving anyway.'

'I'm leaving?'

'Well, you're leaving at some point, I know that much, and I can hang out until then because I am not leaving. Just let me know when you've got your bus ticket, I will be over there.'

I was helping out a couple of other places, I was going climbing again, making use of just being back in town and not having to worry about working.

That was when I saw the Germans walking down the street and I am like, 'You guys need to fix the floor!'

James was, 'Argh, where is the money going to come from?'

I responded, 'It needs to be done, and these guys are going to do a fantastic job and they work for peanuts.' That was the one thing he queried; he was untrusting of my judgement. 'Dude, trust me, they are going to smash it, let me do my thing on this. Just wait.'

It was the only time I ever put him in his place like that. The Germans absolutely killed it. They did it section by section, we never closed. We never even closed off a section at night and they built us a new bar and it's as sturdy as fuck. They did everything they could and all for food, drinks and accommodation.

When it was done, James got it because it was a mess when I got here, a complete mess. It had a good vibe and it was busy, but there was no way it would be able to handle the amount of people we have got now. Back then it was only half as full as it is now but everyone was super pumped on it.

[A man comes over and talks in Lao asking for money. He and MR SAKURA exchange a few words and MR SAKURA hands over some money.]

MR SAKURA: He didn't do what I wanted him to do and he charged me a lot for it. He's just finished the roof. I want it to be vented better.

ROBSON: It does get hot in there.

MR SAKURA: Yeah. I wanted the top bit, I tried to draw him a picture, I try to speak the Lao, as much as I can but I don't know a lot of the construction terminology. Like how do you say ventilation? And I am a pretty shitty artist so trying to draw and stuff isn't easy. I want it to be like a big fucking hat with a fan sucking all the hot air out.

'No, but the rain is going to get in,' he said.

And I said, 'No, not if you fucking do it right. You're the fucking guy building it, you should know how a vent works.'

Anyway, he charged me a fucking lot, 1,000 dollars in labour or 7.5m kip, for 3.5 days work. Him and two guys. Steep like. I wouldn't pay someone in fucking Canada that. That's one hundred dollars each a day pretty much. That's… yeah, so anyway, he's not doing any more work for us.

He wanted to redo the front for us, it was all wooden. I told him to just clear out all the wood stuff and fill it with gravel. I want to build stuff on top of the gravel but for now, it's ok, it will do, whilst it's raining anyway. He wanted 3m kip to do that, I paid two guys 100,000 kip each.

ROBSON: What is that mural there - Sakura with the dice?

MR SAKURA: They used to have a thing back then, when it was more like a restaurant, where you would roll a dice and if you got a certain number you got a half price drink. That was their thing.

ROBSON: Have you thought about getting some dice? Dice are good. Dice are always a bit debaucherous… a bit, you know…

MR SAKURA: Fuck, yeah, some good games. I used to play this game called the Mexican, it's pretty sweet, if you roll a… yeah, I'll get some dice.

ROBSON: Have you read The Dice Man?

MR SAKURA: No

ROBSON: It's a good book.

MR SAKURA: Yeah?

ROBSON: You'd like it actually, it's about this guy, a psychologist in the 70s and all about Freud and Jung and why both are flawed. Basically, he has a nervous breakdown and decides to live his life by the dice. He makes all his decisions by the dice; what to do, where to go, whether to cheat on his wife, do whatever, you gotta read it, you'd like it.

MR SAKURA: Cool, I like that.

ROBSON: I lived by the dice for a whole summer.

MR SAKURA: Nice, that's fun.

ROBSON: I ended up in Amsterdam.

MR SAKURA: That's a great place, Amsterdam. Hehe, roll a 3…

ROBSON: 'Roll a 3 we're going to Amsterdam. Roll! Alright, looks like… get your passports'

MR SAKURA: Amsterdam is a good one. Yeah, I could certainly get used to playing some dice games.

ROBSON: So what else do you need to do with the bar?

MR SAKURA: We are going to do a beer garden where people can sit down and it's not too hot. I talked to Namkhong, the beer company, and they're gonna make a Sakura Namkhong beer garden. They are going to sponsor it and give us some tables and umbrellas. I will make it nice and chilled so you can dance and party in the main room and chill out back there. I got some things on the go.

ROBSON: Who lives in the house behind where the beer garden will be?

MR SAKURA: Joey's brother lives there. I lived in there for a while, then the air conditioning started breaking in the middle of the hot season. Not cool. Not cool at all. I'm from Canada. I like it chilly.

I still have to sort out the electricity. All the wiring is held together by balloons and stuff like that. I got a pretty good thing going so far. A lot of help from people who used to work at Sundowners and Happy. My one buddy, Hans, did all the menus. He sent it over from Germany. He used to do all our flyers and stuff. He's a graphic designer who is big into Photoshop. It is all just a continuation from what Sundowners used to be and then built up to Happy, and now that built up to Sakura. It's only the name that is slightly different. The whole thing has been a journey.

Every year something would go wrong at the end of the season and that would kind of make the year almost pointless, but we learnt from everything. This time I think we're actually… fingers crossed. This is what we've always wanted to do and the location is much better. We are in the middle of town and people get dropped off from the buses down the street. We are across the street from the Irish Bar. He is a good guy. He's been here pretty much as long as I have. I have always had a good working relationship with him. That's always been good. I don't take his business. I talked to him about it and made sure I wasn't stepping on his toes or being too loud or anything.

'Let me know please because we are buddies.'

He said, 'If anything you're helping my trade because you are bringing more people to this street.'

This road was never that busy, it was always the next one. We're bringing people to this street and everyone is getting a piece of it.

Joey is the business owner, he does all the financial stuff. He's doing alright. Finally. For the longest time, it was a running joke that we always get shafted at the end of the year. I'd say to him, 'Joey you always have a pretty good bar but you never have any money.'

ROBSON: What does he spend it on?

MR SAKURA: He spends it on improving the place and like, yeah, just business ventures that didn't quite work out. We tried throwing a couple of festivals. We booked a super popular Thai Reggae band to come up here and then on the night a big storm hit so no one came. We didn't make any money and it cost a fortune.

Every year there was something that fucked up and we were left back at square one. Every year, it's always after his birthday, when something always happens. There's always something. We made it through his birthday this year. This year, [he smiles] this is the longest we have gone without something happening, I think that's why he took off for the Philippines for the last month to have a holiday.

'Joey, just trust me 100%, let me do what I want to do with the place.' I'm a lot more cautious this time round and that is why it's working, it's really paying off.

ROBSON: You've grown up?

MR SAKURA: Yeah, I think so.

ROBSON: Reckless in your twenties maybe, pushing it too far?

MR SAKURA: Yah, yeah, humbling experiences…

ROBSON: What did you think you were going to be when you grew up?

MR SAKURA: The first time someone asked, the first time I really thought about it, I said a dentist. When I was 13 or something like that. I had to plan courses for junior high, so I went down the science route and thought I would be a dentist and make some good money. My dad was a computer programmer and my mum managed a home building company doing property stuff. At junior high, I discovered I didn't really like science that much, haha, at high school even less. I was always part of the student council; president or vice president, always into people stuff. You know? I like being in charge. I like being the boss.

CHAPTER FORTY-TWO

MR SAKURA ON LOVE

ROBSON: Do you fall in love easily?

MR SAKURA: Hmm, no.

ROBSON: Probably not if you've got the J in Myers-Briggs. It is very hard for the Js to fall in love. Have you ever been in Love?

MR SAKURA: Maybe twice… in my life. All from here but then they've moved on.

ROBSON: So for them, it is a fling with the bar owner and for you, you're thinking, 'Oh my God, you're the woman of my dreams!'
They are thinking, 'I've gotta be in Don Det tomorrow.'

MR SAKURA: Yeah, or 'I gotta go be a dentist…' That one… she was pretty long term.

ROBSON: Did she stay here to be with you a while?

MR SAKURA: She was here for a few months. That was a good one. She had to go back to the States, she was from North Carolina. She was fucking awesome. I went and visited her when I went back to North America. I got to meet her family, that was strange, they were a really traditional Italian family. They were very proper with lots of good money, a little intimidating.

They were asking me, 'So what are you doing with your life?' That wasn't awkward at all. I at least did have some sort of story though because at that point I was going back to Canada to pursue being a teacher.

Actually, I did have a story to tell, 'I've been doing tourism in Lao, I have got my degree, I'm going to be a teacher and stuff.' It didn't quite work out quite like that but it was a good enough story to keep them happy for long enough. I spent like two weeks out in North Carolina and she came up to Canada for Christmas.

It was really tough because I said, 'I'm not going to keep going down to North Carolina and you're not going to keep coming up to Canada. You should finish your schooling and be a dentist. You know where I will be.'

ROBSON: Do you think she'll come back?

MR SAKURA: Probably not. It's ok, I got over that.

ROBSON: That's hard.

MR SAKURA: Yeah it's tough right? Especially when you are used to not really caring. You know? Normally when people leave I am just, 'Cool, whatever.' Probably glad, 'It was kind of awkward, waiting for you to go, you kinda overstayed your welcome.'

But hey, I know if they're cool enough, if I really like them, I will see them again in the future, just like my friends. I met some amazing people out here, and fuck, I will always be here, and if they leave, 'Oh bye, see ya next year or in a couple years.' I will be going to Europe next summer, I am saving up for it now and yeah there's a tonne of people I will get to see again. It will be sweet, it will be cool. Hopefully, they won't be married and have kids by then. It's fine - I am ok with my choices in life.

Sometimes I fuck it up, right? Sometimes I really like a girl and it doesn't end up working. Actually, when I went to Goa, I was talking with this girl and she was like, 'I am going to Goa at Christmas.'

'Sweet, you want some company?'

We were hooking up for a while out here before that, but it just wasn't the same when I went there. It got a little awkward.

ROBSON: How long were you in Goa?

MR SAKURA: Two or three weeks, something like that, just on my holiday from teaching in Vientiane; a Christmas holiday.

ROBSON: High season, good time to go. Good parties, do you like Goa? You'd like Goa.

MR SAKURA: Goa is nice, it was cool. We went all over it, pretty much stopped everywhere we could. We were also with a couple of her friends which was just weird. I thought we would pick up from right where we left off and just hook up.

She just wasn't into it. She wanted to go see stuff and I just wanted to see her… for her she was more on vacation, she was there to see Goa.

ROBSON: So was she the other one you were in love with?

MR SAKURA: I wouldn't say love but definitely up there. When she said, 'It's not the same, we're not going to be doing anything,' I was a little hurt.

'What the fuck!!?? Fuck me for two weeks… it was fine in Vang Vieng. Well, there goes those plans,' it was just frustrating, 'Why didn't you tell me?'

'I thought we were just friends,'

'Oh fuck off…'

ROBSON: Did you stay with her the whole time or did you just ditch her after that?

MR SAKURA: We were on the same kind of path so we would just be partying together up and down the coast. She was hanging out with other dudes and I was getting jealous, and I don't like to be jealous. 'Fine!' so I just did a bunch of MDMA. It was still fun and I still got to see lots of cool stuff but that wasn't exactly what I went to Goa for.

Oh, the American girl… she was the one.

ROBSON: She was the one? Maybe, maybe in the future when the stars align…

MR SAKURA: I know she really liked it out here. We still keep in touch.

ROBSON: She was out here for how long?

MR SAKURA: Three months.

ROBSON: It is not unfathomable that she would live here then?

MR SAKURA: Maybe. She was my girlfriend when she was here. She is super fun, she likes to party, likes festivals. It was just not the best timing which I can deal with. If it happens it happens and if not it was fun whilst it lasted.

ROBSON: Because you have got things to do – you're a busy guy. This is what I was saying, it's hard for the Js to fall in love.

MR SAKURA: Yeah.

MR SAKURA: See the Js like to have a lot of structure, which is interesting because love is not a structured emotion. It comes out of nowhere and you have to roll with it. For a J, they're organised doing their thing in an ordered structured way. Love comes along and they are, 'What is this? Get out of my face, I am busy right now. I like you but…'

MR SAKURA: Yeah, '…I already had these plans and I'm doing my shit.' Yeah, exactly like that.

ROBSON: That is why I am asking you as a J, you probably find it harder to fall in love or adapt around love.

MR SAKURA: Oh yeah, it is definitely harder to fall in love.

ROBSON: It's just not a priority, not on your list…

MR SAKURA: Yeah, yeah… it was on my list when I went to North America… but it didn't happen.

ROBSON: When you went back you thought maybe you were closer to her?

MR SAKURA: Yeah, I am closer to her, maybe we can work something out while I am going to school to be a teacher and she's doing her thing. We could finish around the same time and whatever, you know, hopefully, thinking… but I love this place more.

ROBSON: I was about to say if you really loved her more than anything you'd just move to North Carolina?

MR SAKURA: Yeah, exactly. I am not moving to North Carolina.

We had the discussion when she came up to Canada, 'Ok, it's not the time.' I don't think I let her down, I think she probably wanted to experience college life.

ROBSON: It must have been so hard.

MR SAKURA: [Pensively] Yeah, yeah.

ROBSON: Well, she probably loves you back…

MR SAKURA: Yeah, I guess so.

ROBSON: And she knows you love her, and how much you love it here, and she probably wouldn't want you to stop doing your thing to move to North Carolina.

'Don't move here for me, that's a lot of pressure on me, you'd be bored as fuck here.'

MR SAKURA: Yeah that's pretty much exactly how the conversation went. Haha. She is going to be a student and smash it and if I am there it's also a distraction, because what the fuck else am I gonna do?

ROBSON: She will be the only person you know there. Ahhhh I feel the pain, I feel the pain for you.

MR SAKURA: Yeah it's true, a rough one but… c'est la vie. I've got things to do.

CHAPTER FORTY-THREE

MR SAKURA ON THE FUTURE OF VANG VIENG

MR SAKURA: That's what I really want; for Vang Vieng to grow as a town. With more actual stuff to do and to be a little bit more diverse. A tourist base. Any way I can help or support someone or help this I do. A lot of the Lao people aren't really too sure what foreigners or tourists are that into, so they end up trying to do what everyone else is doing.

I spent a lot of time in tourist towns like Banff in Canada, for example, and before I came here I spent a lot of time in Cambodia and in Thailand for a bunch of months.

ROBSON: Did you always have visions of adventure tourism or is that more of a new phenomenon?

MR SAKURA: That was more a result of the bottom falling out of Vang Vieng and the shut down because now we need new reasons for people to be here. They can't just come for the alternative stuff now.

Whether people are doing that or not, it's still a little shallow. If there is no tubing there's not a lot to do, what is drawing people? And there are loads of things to do here. For starters, the place is just incredibly beautiful. You can go climbing, caving, kayaking, trekking, hire bikes and All-Terrain Vehicles, quad bikes. The Blue lagoon is pretty awesome too. We just need more people to come here to visit and more people to come and set stuff up. There is such a big opportunity here.

It has been a focus of the government or tourist board, trying to promote adventure tourism and other stuff to do. They have been focussing on the Asian markets like China, Japan and Korea.

We also lease the place next to us, so we want to knock down that wall and make Sakura into a super club. We want to keep growing as much as possible. We use as much of our profits as possible to reinvest. I am all about that, which is why I live in such a shitty room. I am not used to spending money. I grew up pretty broke and have always been kinda broke. I don't even buy new clothes and shit like that.

Amazing people come through here. Some of the most incredible promoters or bar staff or positive people I have ever met. Just really encouraging, there's always someone who is real amped up to be here and real excited about everything. It makes it real hard to calm down. Fuck, there is always someone who is like, 'Stephen you could have this place way busier, let's fucking do this…'

I'd respond something like, 'Uh, yeah, of course, I was just gonna go to bed, but oh yeah, yeah… especially with Sundowners. That was the very beginning of it. Just getting people to find it was pretty difficult so everyone wanted to help get it going. There are just a tonne of amazing people coming through VV.

That is probably my favourite thing about this place; meeting the people. Locals as well, they are the most generous, genuine, awesome people. Let me do whatever and everyone is really digging it, but the other tourists that stay, the staff, are real great now. They've been around for like forever now, and all the Lao guys, it's the same team, same people from Happy. Sundowners was Joey's brother, his wife and a few other foreigners who passed through and ended up staying.

I have been accumulating people and getting people to stick around a lot longer than they originally planned to.

ROBSON: How do you do that?

MR SAKURA: I am just always as positive as I can be. I have no idea maybe just magnetism? How do I do that?

I think people just always want to be around that. It is also a really shady bus ride out of here, whichever way you go… 'You could go to Luang Prabang for eight hours… or you could just stay another day.'

ROBSON: Do you want more Sakuras?

MR SAKURA: I am not too sure what is next. We need a guesthouse because everyone needs to stay at a guesthouse. And I'd make it a fun one. A cool one, with a chill-out area, a pool table and in a really beautiful spot. Somewhere to lounge around. You'd have your own tuk-tuk driver.

I have wanted to have something like that for quite a while. It would be really cool because you get to know everyone else who is staying there and it would just really be bringing people together. It would have a really cool vibe and cool people and you would meet really cool people. We would have really good food there. It would serve as a link to the bar, maybe offer two free beers at Sakura or something like that.

Also, we have been donating money to schools. I want to do a bit more stuff like that. Maybe organise some volunteer teaching. That is easier to do out of a guesthouse than a bar… it doesn't have the same kind of respectability. 'You can also go teach at this school… you bunch of drunks…' I am not sure that is the right type of thing. I've been wanting to do something along those lines. There isn't really a good English school in Double V.

Joey's partner is managing EEFA - Equal Education For All. It is a non-profit getting volunteers to teach English classes after school to Lao children. He's linking up with the bar, and some people working here go and volunteer there. He's really passionate about EEFA and anything we can do to help promote the organisation will be good. It is a great cause.

To find out more and to volunteer please visit: **www.eefalaos.org**

To donate visit: **www.justgiving.com/crowdfunding/robsondob**

I would like to teach more people English and even teach things like business and tourism management. I really want everyone to chip in, I don't want to run everything myself. I want everyone who lives here to benefit and learn how business models work when dealing with tourism. Teach everyone how to attract people to their businesses. Everyone just copies everything else and it is exactly the same thing everywhere. I want diversity, I want people doing different things.

Then we can start offering combo packages, anything that gets people to stay longer. If we give them more to do, different activities and if the local people are more savvy, that's when we can really put VV on the map as a credible tourist destination.

That is what I like about Cambodia. The locals have a lot of business sense. The kids are on it, they're selling you stuff, they're funny, they are entrepreneurial. They have to because they are so poor. Desperation breeds ingenuity. Here, it's part of the charm that everyone is really chilled and lazy and you can walk into a shop and wake someone up to buy something which is neat, but it would be nice to have a middle ground.

It is not a money making culture. The weather patterns here are quite predictable. There is always something growing and everyone has got food. As long as everyone is doing a little bit everyone can eat. If there is food on the table and beer in the fridge then they are pretty pleased with that.

It is slightly different in Vientiane where there is a bit more striving. It is weird there because they all want almost the same thing. Everyone wants to drive a big SUV. Everything is really superficial. They will have a massive house and when you go inside there is not a piece of furniture. It's not comfortable, it's not cosy. I'd much rather be in a smaller house with a nice sofa.

There, all the money is on the appearance. 'Look at my big house.' It's very apparent in Vientiane. A lot of them won't have too much money in banks either, it's all in gold and they would always be wearing it. If you see any rich people they will have big gold pendants. There is a goldsmith in every little town or silver if they aren't as rich.

ROBSON: They don't get mugged?

MR SAKURA: I don't think so. It is not like that because of the punishment and the culture. It is deeply Buddhist. Buddhism is very much embedded in the culture.

ROBSON: There's a guy who owns a guesthouse you should meet. Right down by the river. I will take you down there if you like. That is where I stayed the first time. I had a coffee with him and chatted for a couple of hours. He used to work in agriculture then he retired early and bought a plot of land by the river. He set up a guesthouse. It started as bungalows and then he's had four iterations of it and now it's concrete. It is beautiful and the view is awesome. He said through his work for agriculture he had been everywhere in Lao and this was the most beautiful spot he could find. Since Lao is one of the most beautiful countries in the world you could almost argue that this is one of the most beautiful spots in the world.

MR SAKURA: I can agree with that.

ROBSON: He'll be a good ally for you in your mission of making this into a tourist place because he kind of wants to do the same. He is in the industry, he is on the guesthouse association.

MR SAKURA: I am sure I know him.

ROBSON: You must have met him if you have been here six years but he didn't know Sakura Bar ironically. It's so funny because you live three minutes away from each other, because I said to him, 'What about Sakura Bar?'

He was, 'Hmm, I think I walked past there once.'

MR SAKURA: It is a different village. There are like seven different villages making up VV. It is funny, it seems like it is one big town, but every village is centred around a temple. So you have the temples just over there, another one just past it and then one over here and one over there. At all, like, the village festivals you interact with people from that village. They don't really come into the next one other than for shopping and even then you'd go to the market and that is far way.

I have lived in a couple of the different villages. I bumped into someone at the market the other day whilst I was speaking Lao and he said, 'Why do you speak Lao?'

I said, 'Because I have been living in Vang Vieng.'

'Wow, which village?'

'Uh, like, Vang Vieng village?'

'I would know you, I used to be there but now I am in Savang village.'

That was when I realised there are lots of small villages. So, for example, Sundowners was in Vang Vieng village and Happy Bar, forty metres away was in Ban Savang. The bridge separated the two villages or the street that led up to the bridge.

It is an interesting time of year around March when all the villages have their own festivals. They will never be on the same day. One village's festival on a given day and everyone in the village opens their house and has cases of beer and food. You just walk around and everyone is, 'Come, come sit down.' You can cruise around all day and the next weekend it will be the next village's festival. Walk around, sit down, chat and laugh with the locals. It would be completely different people because it is a different village.

I didn't stick to one, but I am quite unique, a white man in Lao. I speak Lao. Not many white people speak Lao. Some speak Thai but few speak Lao. They are slightly different, maybe like Portuguese and Spanish. You can understand each other but they are different.

Speaking Lao is definitely fucking awesome. Lao people are really fucking funny. They are very carefree; they like to joke about shit all the time. They don't take too much too seriously. The ones that deal with tourists or foreigners a lot are a bit more edgy. They are a bit more jaded as they have to deal with a lot of shit, but a lot of ones that don't speak a lot of English, they are fucking hilarious.

The Lao are cool. A lot of them came from other towns and moved here to work in the tourism industry because there is money to be made. For those that are a little entrepreneurial, they come to VV. A lot of our staff are from Luang Prabang or Vientiane and they speak a little English so they are probably quite educated. I would love to have just Lao people working here, that would be great. The problem is I still need foreigners to draw in foreigners. I treat both equally in terms of payment. Lao people get paid more here than anywhere in town.

This is good for the community; generating revenue, paying locals, donating to school. We have government certificates for doing the school work. I just try to help as many businesses as I can. There's one climbing place – I got all of their climbing equipment for them from Canada. Just filled up my bag on a couple of trips back.

'Hey, you don't have any proper ropes, why don't I get some for you?'

'Really?'

'Why not, I am coming here with an empty bag anyway.'

So I did. That is one example of that; just trying to help everyone and make it a better place. I just want to do more of that. The people here recognise it. In Vancouver, you could be the nicest guy in the world and no one picks up on it. No one gives a fuck.

CHAPTER FORTY-FOUR

THE COMMUNITY

ROBSON: I can see it all coming together, you know, like with the pizza place.

MR SAKURA: Yeah.

ROBSON: You've worked in kitchens, you like bars, you worked in bars.

MR SAKURA: There are a lot of bars, Sakura is an amalgamation of all the bars I have been to around the world. Every good bar I have been to, I pick up on what is working, how things flow and how people pick things on a menu. 'Price wise, how do I get the best deal or the most bang for my buck? If I only get two shots it's only 5000 kip.' All the backpackers are trying to do the math in their head, how to get the drunkest for cheapest. Koreans just don't care, they just spend the money which is awesome.

ROBSON: I think we need to translate this into Korean.

MR SAKURA: Oh yeah, yeah, totally. It will probably be our biggest market.

ROBSON: I think so.

MR SAKURA: We need a Korean author or biographer or whatever.

ROBSON: You probably know more Koreans than I do?

MR SAKURA: That's true and that's another thing. There is this one Korean guesthouse, Blue something or whatever, and he has the Korean BBQ too. Him and his brother always come in and, 'Fuck, I seen you in here for a couple months now, what's up? Most Koreans are only here for a few days.'

He told me he has this restaurant so I now go there once a week and have a Korean BBQ. They feed us for free. We give them free balloons. I think they're just pumped. They are just slaying the Korean ladies because they just come in high fiving all the staff and there are these new girls every time and they're, 'Huh… we feed you once a week and we are getting laid regularly.' That is how I am taking that. They are helping bring people in for sure.

ROBSON: That's good.

MR SAKURA: I want to get as many people working together as possible.

ROBSON: I find it fascinating the connections you've got. The Italian guys with the pizza place, the Korean guys with the guesthouse. It is an odd combination, they may not have met independently but they would meet through you and there could be a lot of cross-pollinating maybe with the Koreans and the Italians. You're connecting it all together.

MR SAKURA: I told the Italians, 'If you get a night off, let us know, we'll take you to the Korean place and eat some of that.' Same with the Korean guys, 'I'll take you for a pizza.'

ROBSON: So the expat community here?

MR SAKURA: Fledgeling.

ROBSON: You have the random people passing through who stay and then those who are setting up businesses?

MR SAKURA: Yeah, and those ones, there's a handful, there's the Irish guy at the Irish bar, the Australian guy at Oz bar.

ROBSON: Who we met yesterday?

MR SAKURA: No that's another Aussie who works there. They are next to a Norwegian, he's got a guesthouse in a beautiful little spot, and then there's a Brit, who's got a hotel, he has been here the longest. He is the only survivor of the first purge fifteen years ago. There's a New Zealander who's got a great hostel and he has been here for fucking ever.

The older expats, there are four of them. Then the Irish guy and I showed up at the same time. The Aussie guy he is older, he kind of showed up a couple years after I did. Even then it's still tiny. What is great is that you get an international expat community helping each other out. A lot of places like in Vientiane, your different expats from countries will all hang out, like the Brits, the Germans the French will all hang out with their own and they will never really mix. If you get it early enough everyone can hang out together.

ROBSON: That is what it is when you talk about community or is it more with the Lao people?

MR SAKURA: It started with the Lao people but I am trying to get everyone together. I just want a welcoming attitude because that is very Lao or at least that is what they want to promote.

[A man comes and sits down next to us.]

MR SAKURA: This is Joey's brother. He used to run Sundowners.

ROBSON: What do you think about Mr Sakura?

MR SAKURA: What do you think about me? [Translates it into Lao]

JOEY'S BROTHER: [Speaks in Lao] Lav pen nongsai khongkhaphachao

MR SAKURA: Like his little brother I think.

ROBSON: He thinks of you like you're his little brother?

MR SAKURA: That is what he said.

ROBSON: Amazing.

MR SAKURA: [Smiling] Little brother… that's a good one.

CHAPTER FORTY-FIVE

DRINK TRIPLE

ROBSON: What does Vang Vieng mean? Someone told me it means 'float on' or 'keep floating'…

MR SAKURA: I don't think so, I don't think it has any meaning. I don't know.

ROBSON: Is this Vang Vieng where we are now?

MR SAKURA: No, Ban Savang.

ROBSON: Vang Savang?

MR SAKURA: Ban Savang.

ROBSON: Ban?

MR SAKURA: Ban. Ban is village, Ban Savang and Ban Vang Vieng.

ROBSON: So where is Vang Vieng?

MR SAKURA: Two or three streets over. There are seven, or maybe even up to nine villages that are all considered Vang Vieng. It's like asking someone what neighbourhood you're from.

ROBSON: I guess, I am from London but specifically I lived in Clapham.

MR SAKURA: Yeah, yeah… wait you lived in Clapham?

ROBSON: You know it? You've been there?

MR SAKURA: Yeah, Clapham Junction. I partied down there.

ROBSON: I lived close to there.

MR SAKURA: There's a real sweet bar… I can't remember the name.

ROBSON: What type of bar?

MR SAKURA: Um, cocktail?

ROBSON: Be at One?

MR SAKURA: Yes!

ROBSON: I am in there a lot. That's a really good time bar. There is something about it because everyone is having cocktails and also the interaction with the bar staff is good. The bar staff are really well trained. I think they naturally hire fun people. They do a great Happy Hour too. Are we going to do some type of drink promotion for the book?

MR SAKURA: Maybe 'Drink Triple'? I never really did anything with the 'Triple'. You know the T-shirt says, 'Drink Triple' but there isn't anything in the bar that is 'Drink Triple'.

ROBSON: I've got it. At Be at One in Clapham, we drink the Trinity; it's an Espresso Martini, a Porn Star Martini and a Rum Runner to finish. Let's do that! In Vang Vieng, the Trinity can be The Triple.

MR SAKURA: Oh yeah? No shit? Sweet. That's awesome!

ROBSON: I will tell you the full coincidence or fate story one day, well, you will read it in the book but that has been our catchphrase of the entire trip. And this is all coincidence.

MR SAKURA: Everything, yeah, fucking alright… well, maybe fate?

ROBSON: Who knows… thank you.

MR SAKURA: No thank you, I have really enjoyed it.

ROBSON: I haven't done your personality profile either. We will do it later. Will there be an after party after the DJ thing?

MR SAKURA: Yeah sure. Definitely. There's always an afterparty.

PART SEVEN

DANCE FLOOR

CHAPTER FORTY-SIX

SUPERSTAR DJ

'Alright, ladies and gentlemen. Five more minutes left of the balloon happy hour. They will be a third off at the bargain price of only 10,000 kip. Right now we have a very special guest DJ for you. All the way from London and Ibiza via Las Vegas, it's Robson!'

That night for the first time in 12 years I played a live DJ set out. It was a success. We had chosen each song, with much debate, to be played at different points to create the perfect atmosphere. Every song specially selected and placed to drop in for maximum impact on the dance floor.

I got into it and it was like I had never been gone. The crowd really went for it. The Koreans went crazy first, of course, but everyone followed soon after and the whole dance floor was jumping. Mr Sakura was pleased. Partly because his customers were enjoying the music but also because his staff were really getting into it too. That meant everything to him, the fact that his staff were happy. He is a great leader and has inspired a great team.

The Geordies had a whale of a time too. They were enjoying dancing to the set as much as I was enjoying playing it. One thing that I hate about many DJs is that a lot of the time they look too cool for school. Quite frankly they can look absolutely miserable. They would certainly have the 'I' in Myers-Briggs. In some ways, you have to be introverted to really spend that level of time crafting a mix. DJs are introverts.

Whilst I appreciate it can be highly technical and a lot of people don't realise how much concentration is involved, there's no reason not to be smiling. The DJ is, after all, controlling the beat of the party. I believe you need to have some fun! Put on a show! You are trying to make everyone else have fun after all so you should be having fun too. As a result, I put a lot of energy into my DJing and I told the Geordies we need to personify that energy too. We were wearing our heavily-sought-after white with pink trim Sakura tops and looked the part. We raved in the DJ booth just as much as the Koreans were raving on the pool table. That energy transforms a DJ set and everyone on the dance floor felt it too. It is the same with a rock n roll band. If they're having fun on stage the crowd has fun too.

The pool table was at its fullest with every nationality dancing alongside one another. Peace, love and unity were felt throughout Sakura that night. It was a good old fashioned rave and we had the dance floor gyrating.

I told Mr Sakura to get on the microphone to do his goodbye wrap up.

'But it's ten minutes early.'

'I know, but that will leave some time for the encores.'

He looked sceptical. 'Ok, but I'm telling you, they never ask for an encore.' He got on the mike and even he was taken aback by the crowd's reaction.

'One more song, one more song, one more song…' Geordie Sam was on the pool table encouraging the crowd and they were getting stirred into a frenzy.

'Better give it to them then,' Mr Sakura exclaimed in disbelief.

I finished my set with a series of encores. My grand finale was a little-known remix of Coldplay's 'The Scientist'. A gentle nod to Eliza and Beef and that day we cycled on Don Det. I had had such a good time in Lao.

'I just wanted to go back to the start.'

I dropped in the tune and left the DJ booth to head down onto the dance floor. Various ravers and bar staff alike hugged me, high-fived me, congratulated me and said thank you for the set. I thanked them back and got on the pool table, one last time, with everyone else for the final breakdown.

HGL came up to me. She had two vodka drinks in hand. She gave me a big hug and said, 'I am just having the greatest time ever. I am so happy!' She gave me one of the vodkas and the accompanying Sakura T-shirt that she had got with her two drinks. It was the blue one.

EPILOGUE

COINCIDENCE OR FATE?

The bar slowly cleared out as I basked in the afterglow of my set. I was exhausted but the adrenaline was still pumping through my body. People had stayed, people had danced and most importantly I hadn't let Mr Sakura down. It was all very well him letting me DJ but if his business suffered as a result, I'm not sure I could have forgiven myself. He was happy. And if he was happy, I was happy.

We sat down for a final chat to make sure I had all the facts about him now that I was fully committed to writing the book.

'Stephen Sakura… he is a revolutionary!'

'No, he's not!'

'I think you are, and here's why. You represent breaking free from the norm. You know, you were doing the normal thing in Canada like everyone else. Working in the corporate world and you decided to try for something different. You packed it all in and came to one of the most beautiful spots in the world. You ingratiated yourself with the people here, became part of the family and are inspiring a community. You're helping out with the climbing school, the teaching English programme at EEFA, the business lessons. You have vision for the town, you've put it on the map and you're helping it grow… and above all that, you created the greatest party on the planet. The 30 nationalities, the free drinks, the balloons, the beer pong, the layout, the pool table… and it's all tied together by this T-shirt that has reached a cult-like status. They can be found far and wide and everyone travelling across Asia is wearing one!'

'I guess…'

'Oh wait, but hang on a minute… am I right in thinking they started doing the Sakura T-shirts before you got here?'

'Yeah, that was Joey's partner, he got the flower. The T-shirt itself is the same design Hans and Chula came up with at Happy Bar. The only difference is it says Sakura and it has the flower instead of the Nirvana smiley face. Other than that it's the same thing.'

'You didn't create the T-shirt?'

Stephen shook his head, 'I didn't create it.'

'No way, this whole book is about the T-shirt!'

'You've gotta write a book about Joey and his partner then. Because I didn't find that flower on the internet.'

I tried to remember what else he'd told me over the past two days of solid interviewing. 'And you didn't come up with the catchphrase - Drink Triple, See Double, Act Single?'

'No, that was Chula, and even then he stole it from some ski resort.'

'But it was your idea to come up with T-shirts originally?'

'Actually, I stole that from Angkor What? Bar in Cambodia.'

'And you didn't even come up with the name, Sakura, because it was already called that when the Chinese had it.'

'I just adopted it,' said Stephen.

'But you brought it all together. You connected it. You're the connector. I forgot to tell you, your Myers-Briggs personality type, the ESFJ, is called The Connector. You made this whole thing happen.'

He paused for a moment reflectively.

'I collect cool people. I have been collecting cool people for some time now. That is why when a lot of people ask me what my plan is, I just tell them I am going to collect as many amazing people as I can. Characters who want to be a part of it who don't have anything else to do. Or misfits. Or they don't fit in at home, or they don't want to be part of western bullshit. Whatever their thing is, I just want to keep collecting cool people until I have the coolest fucking town ever.'

'Vang Vieng,' I nodded.

'And Double V is the place to do it. It's perfect.'

'Why do you think that is?'

'Look around… this place is beautiful. The limestone karsts, the valley and the river. The scenery is so untouched and beautiful. There is just fresh air and undiscovered nature. It's a little out the way so you have to make a little effort to get here. The people are amazingly welcoming and as long as you stick to cultural norms, and don't get too crazy, there are a lot of possibilities. The country is opening up to a lot of foreign investment and foreign people. That said I am just going to help everyone and continue collecting lots of cool people.

'Lots of quality people, that is what I need. A couple of girls want to open up a proper hair salon. Some actual massage therapists are going to come and teach Lao people how to give a good massage. We are going to improve the quality of the place by bringing in quality people. Bob in Newquay and Ben plugging away in Amsterdam both have their own ideas. They want to be part of it in the future. We have Craig and Hans who will be back. Chula will be back. All these parts of the puzzle will eventually be back here. Hopefully… or it's just going to be me, sitting on the corner saying, "Hey guys! Oh wait, it's just me."

'So yeah, "Surprise! I didn't make the T-shirt…" But I don't think it would have happened without me.'

'Coincidence or fate?'

Mr Sakura smiled.

'Maybe a little bit of both.'

THANK YOU

Hi Again,

Thank you for taking the time to read this book. I really hope you enjoyed it and are inspired to visit Lao, embark on your own creative journey or go on a hundred day weekend; be it travelling, writing, music, climbing, opening a bar or whatever else your energy might take you to.

If you liked the book, please head over to Amazon, or any other review site, and give it a five-star rating. Then please tell anyone you know who you think might like it too! That really would be most appreciated.

Now you've finished reading 'Sakura: Coincidence or Fate?' you are probably wondering what to do next with your life. Well, Mr Sakura would love to see you in Vang Vieng, so do go and say hello. He'll be sure to give you the 'Vang Vieng Welcome'. I am sure he'd love you even more if you could 'stay longer' and help him; set up yoga studios, cool cafes, restaurants, eco tours, adventure tourism, teach English... and whatever else your imagination comes up with... as well as his next secret project!

We want to create a community around Sakura, both in Lao and around the world, so if you want to be involved get in touch.

And what now of Robson Dob? I am taking my Sakura DJ Show on the road together with glowsticks, Haribo Candy Showers, inflatables and many, many more other surprises (see Bonus Features - Encore). So if you want Robson Dob to DJ your party, festival or club night then use the contact details below.

I am also now working on my second book which is yet another unique and intriguing story. Follow me on social media or email me to join the mailing list and be kept up to date on this and all the other exciting projects.

If you can help publish this and get it into mainstream book stores or make this book into a movie with Brad Pitt and Johnny Depp let's talk!

Finally, if you have, or know of, an interesting story worth telling, you might just be my next book project and I am available for freelance writing.

In the meantime, have a shotgun, say the prayer and remember, if in doubt:

Drink Triple, See Double, Act Single.

As they say in Vang Vieng;
Hats off heads,
Hands on hearts,
Praise thee father goose master shotgun up above,
May these beers go down smoothly,
Efficiently,
With minimal spillage,
And absolutely no after sickness,
Three us!
Two us!
Gun us!

Until next time,

Robson
Bay Area, California
Email: **SakuraRobson@gmail.com** Web: **www.RobsonDob.com**
Instagram/Twitter/Facebook: **@RobsonDob #SakuraBook**

Kick off your message with your Myers-Briggs profile, your favourite Friends episode and any winning hangover cures!

So in summary:

Thank you for reading!
Please post a five star review.
Recommend the book to your friends.
Connect via email and social media.
Go to Lao!

And more generally; just like Robson and Mr Sakura did - seek out paradise, unleash your creativity and live your dreams.

BONUS FEATURES

BONUS FEATURES

ENCORE

When I returned home, I was on a huge high after my travels, especially after the success of the DJ set and the opportunity to write this book. Engaging the creative parts of my brain was such an adrenaline rush that I just wanted more of it. I went through one of the most creative periods of time in my life. By day I was working a corporate job, by night I was an author and at the weekends I was a DJ.

I wanted to build on what I had learned in Lao. I studied how the music industry had changed since I had left and I listened to over 1,000 songs. I downloaded new software, I watched all of the great DJs performing and I went clubbing again. I watched the new class as well as the old, back at work on the festival and clubbing circuit. I knew I was going to go back to Sakura and put on an even better show.

When I returned to Sakura the following January with Armstrong, we had it all planned. The setlist was even more epic. The tunes bigger and better, the drops perfectly timed and even more emotional. With that, we wanted to put on a totally immersive show. The guys at Sakura decided to call it a Tight and Bright Party. On the posters and the flyers, it encouraged people to wear neon and get a free shot. The Sakura T-Shirts available that night were all the brightest ones - neon orange, neon green, neon blue and neon yellow. On arrival, everyone was given a glow stick bracelet. We had giant beach balls to throw around and other assorted inflatables like a giant flamingo, a clown fish, a banana and a shark. They were a real hit with the crowd. We also had a giant inflatable world, which we got everyone to sign and mark where they came from. During my favourite song, we did the now world famous Haribo Candy Shower. When the breakdown kicks in, myself and all the bar staff threw packets of star mix out into the crowd which was another hit. You could see the surprise and joy on people's faces.

'What's this, Haribo?!'

We went crazy in the booth, just like everyone was going crazy on the dance floor and especially on the pool table. Armstrong went out and took videos and I was taken aback by how much the party was jumping.

He exclaimed, 'I don't think I've ever seen a dance floor quite like it. Not that level of energy. Not that level of good times.'

300 people packed in and the atmosphere was electric.

All the staff were really excited and I was told that they'd had the biggest takings in a single night ever, other than New Year's Eve (which had a 4 a.m. license). It was also great for the bar because it was on a Friday night, when the Jungle Party happens, and many clubbers head to that from 10 p.m. onwards. We kept the dance floor jumping until close at midnight! The party-goers stayed, and they wanted encores. We delighted them with several, pleasing the Koreans as well as the 30 other nationalities that came to party.

Joey was happy and so was Mr Sakura. The Tight and Bright Party was my encore.

QUIZ

DO YOU BELIEVE IN FATE?

*****WARNING*****

Warning, this section contains information that will give the story away. Please read this only after completing the book in full.

****

In putting this book together a number of coincidences (or indeed, were they fate?) happened to me, Mr Sakura and the people around us which conspired to make this whole thing happen. The T-Shirt, The Bar and the Book.

Was it all just a series of coincidences or was it fate? Answer the following questions to decide for yourself. In each of the instances below tick the box as to whether you think it was merely coincidence or if it was in actual fact fate. Then tally up your score at the end to find out if you believe in fate.

Your Turn

DO YOU BELIEVE IN FATE?

1. The Lexpedition recommending Lao as a destination and telling me, 'You will find yourself there.'

____ Coincidence or; ____ Fate

2. Walking down the street in Luang Prabang, seeing everyone wearing the T-shirt and bumping into the German girls who recommended us Vang Vieng.

____ Coincidence or; ____ Fate

3. Having a sense of energy that I was not yet finished with Lao and needed to stay longer.

____ Coincidence or; ____ Fate

4. Armstrong having to go back to London for 'business meetings' (actually a date) meaning we cancelled our trip to the Thai Islands and were at a loss for a couple of days. That gave us the opportunity to go to Vang Vieng.

____ Coincidence or; ____ Fate

5. There only being two seats left on the bus to Vang Vieng on the day we wanted to go.

____ Coincidence or; ____ Fate

6. That same bus having Beef and Eliza on and meeting them.

____ Coincidence or; ____ Fate

7. Having a connection with Beef and not knowing what it was, prompting us to go for a beer together on arrival in Double V.

____ Coincidence or; ____ Fate

8. Beef and I both dating best of friends in Spain, the Russian and the Swede.

____ Coincidence or; ____ Fate

9. Mr Sakura's welcome.

____ Coincidence or; ____ Fate

10. Sakura actually being the greatest party in the world.

____ Coincidence or; ____ Fate

11. Mr Sakura meeting me on the terrace, Alonso being in awe of him, prompting Mr Sakura to tell his stories which gave me the idea that someone needed to write his story.

____ Coincidence or; ____ Fate

12. Sally suggesting, 'I'm travelling for the next year, you write the book.'

____ Coincidence or; ____ Fate

13. Writing the prologue on my phone late into the night and forgetting all about it. Only to find it the next day and remembering it.

____ Coincidence or; ____ Fate

14. My Passport being full, meaning no onward travel to Cambodia and therefore having to return back to the North of Lao rather than continuing with Beef and Eliza.

____ Coincidence or; ____ Fate

15. Meeting the Geordies.

____ Coincidence or; ____ Fate

16. Bumping into Eliza and Beef at the Crazy Gecko.

____ Coincidence or; ____ Fate

17. Meeting Adrian Robson who himself had written a book, thinking to myself, 'If he can write a book, I can'.

____ Coincidence or; ____ Fate

18. The Geordies also buying Adrian Robson's book forming the basis of a connection with them.

____ Coincidence or; ____ Fate

19. The Geordies guessing each other's Myers-Briggs and being exact opposites.

____ Coincidence or; ____ Fate

20. The Geordies' broken arm and airline story.

____ Coincidence or; ____ Fate

21. Geordie Sam being the ideal person to have a dirt bike road trip with (based on Lexpedition's journey) as he was a former youth bike champion.

____ Coincidence or; ____ Fate

22. The Geordies knowing all about the music scene, helping me put together my set and returning with me to Vang Vieng.

____ Coincidence or; ____ Fate

23. The pool table accidentally becoming centrepiece of the bar.

____ Coincidence or; ____ Fate

24. Mr Sakura has a Sakura Moustache without even realising it.

____ Coincidence or; ____ Fate

25. Stephen goes to Asia for the first time thanks to a call out of the blue from Craig Winner.

____ Coincidence or; ____ Fate

26. Stephen sees the Angkor What? Bar T-shirts in Siem Reap.

____ Coincidence or; ____ Fate

27. Stephen asks his university professors if he can get credit by writing an essay whilst travelling.

____ Coincidence or; ____ Fate

28. Stephen's trip to Germany during high school making him realise he can take off and not miss anything; prompting a lifetime of future travel.

____ Coincidence or; ____ Fate

29. Stephen getting tired of backpacking, settling in Ton Sai and deciding to learn to climb.

____ Coincidence or; ____ Fate

30. Sitting in the hot tub, listening to Bob Marley, Stephen decides, 'Fuck this shit… I am going back to Asia.'

____ Coincidence or; ____ Fate

31. Stephen goes to Korea and learns invaluable lessons about Koreans - helping him in future with his Korean trade at Sakura.

____ Coincidence or; ____ Fate

32. Me training hard and not getting to black belt at Tae Kwon Do / Stephen being paid to be a Black Belt with no training at all.

____ Coincidence or; ____ Fate

33. The awkwardness in Ton Sai over the squatted land meaning Stephen had to move on and find somewhere else to climb, taking him to Vang Vieng.

____ Coincidence or; ____ Fate

34. Stephen asks what to do next in his journal at the same time as Joey asks him if he wants to help get the bar busy.

____ Coincidence or; ____ Fate

35. Stephen able to help Joey and his family by using his knowledge of all the bars in the world he had been to. Years getting drunk in places like Banff and across Asia coming in useful!

____ Coincidence or; ____ Fate

36. Stephen helping the family buy things. The House that Stephen built. The family being so kind to him and holding a Baci making him feel part of the family.

____ Coincidence or; ____ Fate

37. Stephen ringing Dean up who shows up out of the blue. This continued the story whilst he returned to Canada.

____ Coincidence or; ____ Fate

38. Dean convincing Stephen that he had to go back to Lao when they were both back in Canada.

____ Coincidence or; ____ Fate

39. Working out that based on his Buddhism essay, his professors would be cool with him doing anthropology essays on his time in Lao. This made it easy to split time between Canada and Lao.

____ Coincidence or; ____ Fate

40. Rainy season fitting perfectly with when Stephen needed to go home to study.

____ Coincidence or; ____ Fate

41. The price of flights dropping on the 6th January, meaning Stephen could have Christmas and New Year in Canada and arrive in Lao just as high season was kicking off.

____ Coincidence or; ____ Fate

42. Chula seeing the 'Drink Triple, See Double, Act Single' catchphrase at a ski resort in Canada.

____ Coincidence or; ____ Fate

43. Running out of money. If he hadn't, he wouldn't have cut the deal with the German apprentices.

____ Coincidence or; ____ Fate

44. If he hadn't met the German carpenters first time round he wouldn't have recognised them the second time and they end up rebuilding Sakura as we know it today.

____ Coincidence or; ____ Fate

45. Mr Sakura wanting to be a dentist when he grew up and falling in love with someone who left him to become a dentist.

____ Coincidence or; ____ Fate

46. Connecting the Italians and the Koreans and creating the community based on the model he had experienced in Banff.

____ Coincidence or; ____ Fate

47. Mr Sakura having been to my local cocktail bar in London. The Trinity becoming the 'Drink Triple…'

____ Coincidence or; ____ Fate

48. The T-shirt not having anything to do with Mr Sakura at all.

____ Coincidence or; ____ Fate

49. But, he connected it all together.

____ Coincidence or; ____ Fate

50. Mr Sakura's Myers-Briggs Profile being 'The Connector'

____ Coincidence or; ____ Fate

TOTALS

COINCIDENCE _____ / 50

FATE _____ / 50

RESULTS

If you scored 50/50 for coincidence, I am afraid you do not believe in fate and think all of this was an uncanny (or even just canny) series of coincidences.

If you scored 1 or more out of 50 for fate, congratulations, you believe in fate!

POLL

Be part of the conversation. On the next page, cross off whether you think Sakura was all coincidence or if it was fate. Feel free to doodle, spruce up the page, give some reasons behind your choice and generally go art crazy. Add any other thoughts you had on the story. Then you can post it to social media and see what others decided too. We can keep a running total and see if it was in fact fate!

Your Turn

SAKURA:

COINCIDENCE

FATE

<delete as appropriate>

 Join the conversation @RobsonDob #SakuraBook

YOUR TURN

UNLEASH YOUR CREATIVITY

As someone who was strong at the Sciences and Mathematics at school I often lived in the shadow of my more creative sister - who, at the age of 12, wrote a 150 page Hercule Poirot style murder mystery novel. No wonder I focused on the numbers!

However, since going to Lao and discovering my own creative energy, I have found out how much fun it is to play around with it. In turn I hope to inspire you to get in touch with yours!

I have left the next few pages blank, each with a theme, for you to fill in with your own pictures, doodles, writings and spontaneous thoughts. Maybe it will lead you to a hot tub epiphany of your own.

If you so wish, take a photo and post to social media with the usual tags (@RobsonDob and #SakuraBook. Maybe someone will see it and help you on your path. Or with the second one at least, Mr Sakura would be sure to take an interest. He might even give a limited edition T-Shirt for some of the best ideas. You can also see what others have posted.

Enjoy, play with it and have some fun.

Your Turn

PARADISE TO ME IS…

Share a photo and be part of the story @RobsonDob #SakuraBook

Your Turn

THE BEST PARTY IN THE WORLD…

Share a photo and be part of the story @RobsonDob #SakuraBook

Your Turn

IF I WAS LIVING MY DREAMS I'D BE…

Share a photo and be part of the story @RobsonDob #SakuraBook

Your Turn

MY FAVOURITE TRAVEL STORY…

Share a photo and be part of the story @RobsonDob #SakuraBook

Your Turn

Have a life changing experience and give something back to the community in beautiful Vang Vieng!

Equal Education For All (EEFA) is a non-profit that began in 2007 to provide educational opportunities for children in Vang Vieng. It achieved Association status (#0151) in 2011.

EEFA currently provides 34 free English classes per week in four Vang Vieng villages. These supplementary classes are taught by local and foreign volunteers on weekday evenings. The students all come to class on their own volition, all see English as key to improving their future job prospects and are always happy to be taught by people from around the world.

In addition to teaching English, EEFA also supports local education by building schools, retrofitting classrooms and providing teaching resources.

If you are looking for a unique and fulfilling experience, why not consider volunteering at one of the schools. Anyone can teach; whether you are an experienced professional or haven't set foot in a classroom in years. All that counts is a passion to help the wonderful students there.

Accommodation is available at the Vang Vieng Organic Farm which offers a variety of options; from dorm beds to private bungalows and even natural mud huts. The restaurant onsite serves organic food, locally grown on the farm. You can even get back to nature and participate in the farm-stay program in the mornings; taking care of the animals, harvesting fruit and getting involved with all the farm activities.

Even if you can't make it to Vang Vieng you can still make a difference:

USD 15 provides a student with teaching materials for one month
USD 50 is enough to restore a whole classroom
USD 250 will grant a scholarship for a student to attend a local college

To donate, volunteer or find out more please visit the websites:

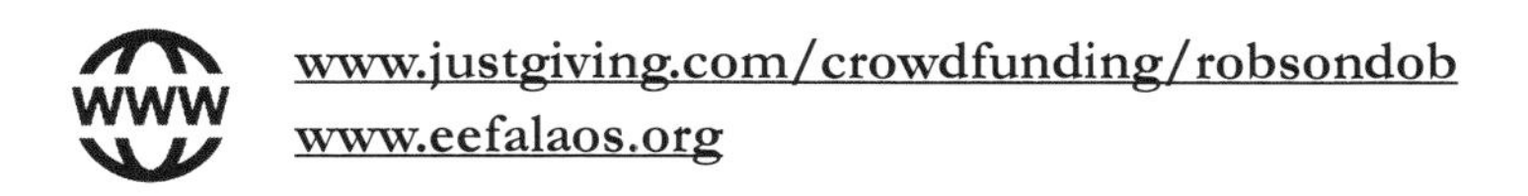

DELETED SCENES

FLIP-FLOP ODYSSEY

Flip-flops. One thing I love about travelling is flip-flops. Your feet feel so free in them. They are the go-to footwear choice of any intrepid traveller. Months on the road spent in this simple but comfortable garment. The Brazilian brand Havaianas is the word's largest and original flip-flop. You probably knew this already, and if you are travelling somewhere reading this, you might even be wearing a pair right now. But did you know they were invented by a Scotsman? A chap by the name of Robert Fraser. He was in Brazil in the fifties when he was inspired by a Japanese sandal, called a zori. They have straw bottoms and are worn by geishas. He recreated them with white rubber soles and, in 1962, was the first to mass produce them as Havaianas (Portuguese for Hawaiian). In fact, to this day the faux straw design on the bottom is a nod to their zori counterparts.

Originally, they featured white insoles with coloured outsoles and straps. They were durable and affordable and appealed to Brazil's working classes and favela dwellers. In 1994, when their popularity had waned and cheaper competitors entered the market, they rebranded. Turning the white soles upside down and creating a number of new monochrome combinations, they were targeted at the premium middle and upper classes. Now worn the world over, by everyone from the Brazilian working classes to Jennifer Aniston and everyone in between, 200 million are sold each year.

Why am I telling you about Havaianas and flip-flops in general? Well, enter Armstrong. On our first night out in Vang Vieng, our infamous night at Bar Fly, we returned to our hotel.

The next morning he opened the door to exclaim, 'Where are my flip-flops?' He looked in dismay to see next to mine were a pair of worn and used knock-off Havaianas that had certainly seen better days.

'Right there,' I responded, in a hungover snooze, 'Right next to mine.'

'No!' he was annoyed, 'I had a brand new pair. These are awful. I must have come home in somebody else's. Begrudgingly, and with no other option, he was forced to wear this sorry excuse for a pair. That was the day we went tubing and experienced Sakura fully for the first time.

The following morning we woke up in an even worse state. Remember the jungle party? Well, we don't! He opened the door of our bungalow on this morning and exclaimed, 'Not again!'

Once more he had come home in the wrong pair of flip-flops. It was even funnier on day two though. Not only were they several sizes too small, they were pink with flowers on them. They were quite clearly a girl's pair. Inexplicably, in his drunken state, he had picked up some poor-now-barefooted girl's pair. They were well made and fairly new Havaiana originals. Again with no other option available, he had to don this pair. His feet hanging over the edges.

When we went for brunch that morning, Eliza excused herself to dash off to buy herself a new pair of flip-flops. Hers hadn't survived the day of tubing and all the fun that had ensued. Armstrong stopped her.

'No need. Have these instead. I was only going to throw them away.' They fit her perfectly.

Fast forward a week and we are sitting by the pool at the Geordie's hostel in Don Det. We were lazing about, having a swim and a Beer Lao. Eliza was sunning herself when Karo, a Polish girl we'd met in Vang Vieng, walked past.

'Who's are these flip-flops?' she scowled.

Eliza looked up meekly, 'Uh…'

'They are my flip-flops… you… you stole them!'

'It wasn't me… uh, Armstrong… Armstrong stole them… accidentally of course… I'm sure he didn't mean to.'

It turned out the flip-flops were a gift from her sister who had visited the Havaiana factory when she had traveled to Brazil and had them custom made for her as a gift.

Jesse J heard the entire conversation and wandered over afterwards, 'Are these "The" flip-flops?'

On the 24-hour bus ride from Vientiane, he had had the misfortune of sitting next to Karo. She had talked non-stop about how someone had stolen her special customised flip-flops and how upset she was about them. She was angry at Armstrong because he had stolen them but she was grateful to Eliza for getting them back. Ironically, if Eliza hadn't lost her own flip-flops she wouldn't have taken the flowery pair Armstrong had found and therefore wouldn't have reunited them with their owner. Luckily she has the same size foot as Karo. Just another one of Lao's coincidences or fate.

OUTTAKES

I had a lot of fun writing this book and Mr Sakura had great fun recounting the stories. Writing a book is hard and his memory isn't all that great. Here's an outtake from our interview to give you an idea of just how difficult it was piecing this all together.

MR SAKURA: But it was… well fuck and I, made all these friends, because it was such a good year. Hey, oh fuck, time to go back to Lao for a good one. A good little stint, a ten month or so. Did I switch it up… trying to think because I don't really remember because I did do a Christmas and New Year's here. Nope, I didn't go back to Canada from London… London… oh yeah, yeah, yeah, that one. Ha, no that was a different fucking trip. It all gets jumbled, um. Ok, so, London, then Germany. Did I go to Amsterdam that year? I don't think so. At some point, I came back here. And uh… January, July, September. I must have spent New Year's here. Was that? No, different one. I don't fucking know. I was here anyway for I think I went home for Christmas on this one because I hadn't ever missed a Christmas at home.

THE OFFICIAL RULES TO

WHEEL OR NO WHEEL?

GAME PLAY

What you need:

1 pack of cards, 2 or more players

Objective of the game:

Gain the highest score from the cards you win a hand with. Low card wins a hand, highest score wins the game.

Instructions:

One player is dealer. He/She shuffles the deck and deals the first person to their left one card face down. The player looks at the card and decides to wheel (keep the card) or no wheel (discard the card) into a discard pile face down.

If the player does not wheel first time they are offered a second card.

Again they have the option to Wheel or No Wheel?

If they do not wheel a second time they are required to automatically wheel (keep) their third card, called a Blind, which is placed face down in front of them. They are not allowed to look at this card until the big reveal at the end.

This process is repeated each time so that all players have a card that they have wheeled.

The dealer then deals herself a card face up. The dealer has two chances to wheel. If she does not, she will go blind (played face down). The big reveal happens as the player to the left then reveals his card and so does everyone else in turn.

The person with the lowest card wins. Aces are low. Note if two or more players have the same value card then they cancel each other out and the next lowest wins. On the rare occasion that all cards cancel each other out then everyone participates in a deciding game of Rock, Paper, Scissors.

The winner of a hand gets to keep their winning card and places it face down to their side. The dealer passes the deck to the player to her left who then becomes the dealer and the process is repeated for as long as agreed. Usually three rounds through the deck but this depends on the number of players and time restrictions. At the end of the allotted time players add up the total value of their winning cards for the final reckoning. Face cards (J, Q, K) count as 10. Highest total wins.

SPECIAL CARDS

The Ace of Spades is the all-conquering card and cannot be cancelled out by another Ace. Winning with an Ace of Spades allows you to steal one of another player's winning cards. Note if it is won with the first hand you cannot steal a card as there are none to steal.

The Seven of Hearts is a magic card. If you win with the Seven of Hearts you can double the value of one of your other winning cards in the final reckoning. It is hard to win with a seven but highly lucrative if you do. Always wheel on the Seven of Hearts.

If a Dealer is blind they are allowed, once every round, to steal another opponents' wheeled card if they so wish before the big reveal. It is worth monitoring who has a blind and who was a confident wheeler. More advanced players may try to bluff when wheeling to hide the strength of their card.

Regional Variation:

In Newcastle, thanks to Geordie Sam, there is an additional special card. The Ten of Halves. This is the Ten of Hearts.

Plays like a five, wins like a ten.

So when wheeling on the ten of halves it is actually a five. But when the final reckoning comes, it counts as ten. This rule adds a new flavour to the game, and worth introducing.

Note: in the world championship and official ranking tournaments this rule is not played.

ETIQUETTE, HAND SIGNALS AND VOCALISATION

At the beginning of every game and once the dealer is ready to deal the first card he/she should state loudly and comically:

'Ladies and Gentlemen, let's play, Wheel or No Wheel?' to which all players clap their hands, stamp their feet, scream and generally make as much noise as possible.

It is customary, every time a card is dealt for the dealer to ask the person, by their nickname, 'Wheel or No Wheel?' If players don't have nicknames, invent some.

The internationally renowned hand gesture for wheeling is pointing an index finger in the air and making a circular motion. Almost like a cowboy waving a lasso.

If a player wishes not to wheel they cross and then uncross their hands and wave them away. Similar to an American football referee gesturing no goal, or a football referee gesturing no penalty play on. No wheel.

If a player is dealt a 'Blind' then all players must cover their eyes and shout 'Bliiiiiiiiiiind'

When a player wins with an Ace of Spades they should play the air guitar and sing the chorus from Motor Head's Ace of Spades.

STRATEGY

Whilst Aces are generally considered the best card in the game, as they are the lowest, be wary of them. You could win four hands with Aces but you will only have four points in the final reckoning and lose out to someone who had won only one hand with a 6 for example.

Beef always used to consider not wheeling Aces, twos and threes as he called these cards breadwinners. He eventually came up with a strategy of only wheeling on face cards and hoping Elisa and myself both cancelled each other out. Whilst a risky strategy it does have huge potential rewards.

One internationally acclaimed player would never look at his cards. Only putting his hand over the face down card and try to feel it's energy before deciding whether to wheel or not to wheel. He was in effect playing blind every time.

Your Turn

Have a go at playing 'Wheel or No Wheel?' with your friends! See if you end up feeling the excitement we did on Don Det.

Post photos of you Wheeling with grace and especially videos showcasing the drama of the big reveal! @RobsonDob #SakuraBook

Perhaps you will come up with some new strategies.

Message Robson if you're struggling with the rules!

Share a video and be part of the story @RobsonDob #SakuraBook

THE TRIPLE

PART ONE: ESPRESSO MARTINI

If you're feeling tired at the end of a long day and want something to pick you up for the night ahead, this is a great one to kick off with. It's made with fresh coffee and will definitely give you a buzz. Hence why it's the first in The Triple.

Invented by the Godfather of Cocktails, Dick Bradsell, in the late eighties at Fred's bar, London. He was asked by a then unknown-but-soon-to-be-world-famous model for a cocktail that was going to, 'Wake me up and then fuck me up.'

The Espresso Martini was his response.

Ingredients:

50ml Vodka
25ml Coffee Liqueur (e.g. Kahlua)
25ml Fresh espresso
Dash Sugar syrup

Instructions:

1. Place crushed ice in a cocktail glass and leave to chill.
2. Add all of the ingredients to a cocktail shaker with plenty of cubed ice. Tip: Add the espresso first so it cools down or alternatively chill a batch of coffee in advance if you plan on making several.
3. Shake vigorously until a white frothy cream forms.
4. Remove the crushed ice from the cocktail glass and pour the cocktail in through a strainer.
5. Garnish with three lucky coffee beans, nestling them on top in the frothy foam.

Robson Recommends:

Experiment with Vanilla Vodka, Spiced Rum or Cafe Padron XO (Coffee flavoured tequila) to give a little variety.

PART TWO: PORN STAR MARTINI

There isn't a cocktail in the world that gets a party started like the Porn Star Martini. It's glamorous! It's brash! And ostentatiously, it comes with a shot of champagne on the side. The sweet vanilla is balanced by the tartness of the passion fruit and complemented perfectly by the bubbles.

Douglas Ankrah, founder of the London Academy of Bartending (read as: knows a thing or two about cocktails) invented this heavenly concoction at the Townhouse in Knightsbridge in 2002. He adapted it from the Maverick Martini, named after a shady club in Cape Town.

Ingredients:

37.5ml Vanilla Vodka
12.5ml Passion Fruit Liqueur (e.g. Passoa)
25ml Passion Fruit Puree
2 Tsp Vanilla Sugar (or dash of vanilla syrup)

1/2 Passion Fruit
1 Shot Champagne (or Prosecco)

Instructions:

1. Place crushed ice in a cocktail glass and leave to chill.
2. Combine the Vodka, Passoa, Puree and Vanilla Sugar into a cocktail shaker with ice and shake.
3. Remove the crushed ice and strain the mixture into the cocktail glass.
4. Garnish with half a passion fruit, floating it on the surface and add a further sprinkle of vanilla sugar.
5. Serve with a shot of champagne on the side.

Robson Recommends:

Originally the intention was to eat the passion fruit, down the champagne and sip the martini. However, no one really knows to do this so I love watching how everyone approaches this cocktail in their own individual way. Some down the champagne, some pour it in with the passion fruit cocktail mixture and some save it to the end. Some people even forget all about it and leave it on the bar or spill it on their way back to their friends, as carrying two or more of these can be fiddly!

Personally, I like to sip the bubbles whilst drinking the rest of the drink and then eat the passion fruit right at the end.

PART THREE: RUM RUNNER

On January 17th 1920, the distribution and consumption of alcohol became illegal across the United States. Within 24 hours the black market for rums, whiskies, brandies and whatever else could be sought became rampant. Rum Running was the illicit activity during the prohibition era of transporting rum into the nation. The Bimini region in the west of the Bahamas became a particularly popular point from which ships would sail to Florida, dropping off Caribbean rum destined for local speakeasies.

In 1972, over a half century later, at the Holiday Isle Resort on Florida's Islamorada, Tiki Bar tender John Ebert concocted the drink of the smuggler's namesake. Ironically, given the name's origins, he had an excess of rum to sell which he needed to get rid of as another [legal] delivery was about to be made. He came up with a heavily rum based drink he could market to locals and tourists alike and the Rum Runner was born.

Fruity and refreshing, it makes an exquisite climax to The Triple. But do not be fooled, its strength is not to be underestimated.

Ingredients:

25ml Light Rum (Originally Bacardi 151 with a punchy 75.5% ABV)
25ml Dark Rum
25ml Banana Liqueur (e.g. Creme de Banane)
25ml Blackberry Liqueur (e.g. Blackberry Brandy or Creme de Mure)
20ml Grenadine
30ml Lime Juice

Instructions:

1. Add all ingredients (except the light rum) to a blender with crushed ice.
2. Blend until all ingredients are mixed and the mixture is thick and slushy, such that you could stand a straw up in it.
3. Pour into a Hurricane Glass or similar. Float the light rum on the top and serve with a straw.

Robson Recommends:

This is my favourite of the three as it is smooth and refreshing but packs a punch. After finishing the Trinity you can do it all over again, although memory loss usually ensues after the 7th cocktail. I just stick to rum runners, thereafter.

Your Turn

Want to have some fun? Why not host a cocktail party or bring the spirit of Sakura to your nights out. You could even start with the Triple and then move on to some of Mr Sakura's favourite drinking games and play some 'Wheel or No Wheel?'. If you're feeling rough the next day you've always got Armstrong's hangover cures to sort you out.

DRINK TRIPLE

SEE DOUBLE

ACT SINGLE

Cheers, Salud, Salute, La-Chaim, Nos, Viva, Bil-afya, Serefe, Kampe, Skoll, Prost, Sante, Kia Ora, Iechyd Da, Skal, Na Zdrowie, Slainte, Kampai, Thum Keo, Gesundheit, Gesondheid, Zum Wohl!

Pop Quiz:

Can you guess the countries?

Please drink the Triple responsibly

 Drink Triple and be part of the story @RobsonDob #SakuraBook

MR SAKURA'S

FAVOURITE DRINKING GAMES

Disclaimer: Please drink responsibly. Do not drive, operate heavy machinery or contact your ex-girlfriend/ex-boyfriend whilst under the influence of alcohol. Abide by your country's laws. These games are for illustrative purposes only. Any actions or consequences that occur as a result of playing these games are the responsibility of the players alone and not of Mr Sakura, Robson Dob, Santa Claus or the Tooth Fairy.

THE MEXICAN

You will need:

2 dice, 2 or more players and plenty of alcohol

Objective of the game:

Rolling two dice to obtain the highest score. The person with the lowest score drinks.

Instructions:

Someone rolls two dice. They can have one, two or three attempts at getting a score. The last throw is what counts, not the best score. The number they decide means every player has that many attempts if they so wish. So for example, if they choose to go with the score on their second roll, then everyone has the option of two rolls. Scoring is as follows:

21 - Mexican (Highest roll and doubles the drinking punishment of the loser)
66 - 600
55 - 500
44 - 400
33 - 300
22 - 200
11 - 100
65, 64, 63, 62, 61
54, 53, 52, 51
43, 42, 41
32
31 - Scumbag (drink and re-roll)

If you roll a 1 or a 2 as one of your dice and have subsequent rolls available to you, then you can keep that die on the table and roll the other one. Thus increasing the chance of a Mexican (or scumbag) on the following roll. If you throw a scumbag you cannot keep the 1 on the table.

The loser is the player with the lowest score. In the event of a tie, the players involved roll again and lowest score loses.

Optional Additions:

41 - All for one and one for all. Nominate one person to drink or nominate everyone to drink.
51 - Nominate one person to drink and every time 51 is rolled that person has to drink until they roll it themselves and then they nominate someone else.

If the dice fall off the table you drink.
When you take the dice out of someone else's hand you have to drink (can be used strategically to trick someone into drinking).

TITANIC

You will need:

A big glass, a shot glass, beer (preferably Beer Lao), soju (Korean spirit). If none available choose another spirit like vodka or rum.

Objective of the game:

Avoid sinking a shot glass of Soju in a sea of beer to ensure someone else drinks

Instructions:

Fill a glass with beer and float a shot glass on the top - this is the Titanic. Players take it in turns to add soju (or equivalent) to the shot glass. They can add as little as a drop or as much as they want as long as they don't sink it. The person who sinks the Titanic i.e. adds the last amount of soju before the shot glass drops has to down the mixture of beer and soju.

Optional Additions:

Loser has to give their best rendition of Celine Dion's My Heart Will Go On.
Girls: If as a result of this game you find yourself floating on a door in the Arctic Ocean, please, please, please, make some space for your lover so that he doesn't freeze to death.

HANGOVER ADVICE

WITH ARMSTRONG

If you want to be a world-class partier than you have to accept the consequences that go along with it; namely the dreaded hangover. Part of the reason why I sought out the enlightenment from yoga and meditation is quite simply because I cannot handle the hangovers anymore. Great for young travellers in their early twenties but harder when you get older. Needless to say, on special occasions, birthdays and visits to Vang Vieng, I do love to party. As does Armstrong. Follow his step by step guide to minimise the effect of your hangover.

Disclaimer: If you are not feeling well consult a doctor. These recommendations are for illustrative purposes only. Armstrong is a seasoned party professional. Don't try this at home. Armstrong is not accountable for your actions.

1. Drink lots of water. When you are partying, when you get home and when you wake up. This is quite simply the most important thing.
2. Have something to eat during or after a session. Wheel a Laughing Cow baguette and that shall be your saviour. Or as Armstrong does, order a baguette with everything on it. Drench it in all the sauces available. Do this with as much enthusiasm and gusto as possible. It doesn't matter if you spray sauce all over your fellow party goers and the woman selling you the baguette - you're Armstrong and you can do as you please.

3. Carb up. Armstrong loves to eat himself into a carb coma the morning following a Sakura odyssey. He has lasagne, pizza, a baguette and beans on toast all in one sitting.
4. Watch Friends. Fortunately in Vang Vieng, there are plenty of places to do this. Armstrong's favourite episodes are:
   i. Series 3 Episode 2: The one where no one's ready
   ii. Series 2 Episode 17: The one where Eddie moves in
   iii. Series 4 Episode 8: The one with Chandler in a box
5. Get fresh air. Again Vang Vieng is ideal and full of natural beauty so it's perfect. But if you are somewhere else in the world, go outside and get some air. Try and do some exercise if you can.
6. Have a rehydration drink before sleep. Armstrong's favourite is an electrolyte sachet you can get from German pharmacies off the shelf or delivered from the internet.
7. Have a fizzy tablet you can add to a glass of water unleashing a month's supply of Vitamin C, B Vitamins, Niacin and other body boosting ingredients.
8. Play David Guetta 'Titanium' followed by Swedish House Mafia 'Greyhound'.
9. Go Kayaking - especially on the Nam Song river in Double V.
10. Have a banana. Any natural sugar in fact.
11. Have a glucose drink.
12. Sleep!
13. Have a cuddle.
14. Get Laid!

Your Turn

MY TOP HANGOVER TIP IS:

Join the conversation @RobsonDob #SakuraBook

WHAT'S YOUR PERSONALITY TYPE?

Answer these four questions. Note down the letter of each of your answers and then put the four letters together.

1. Where do you get your energy from? Externally (E) or Internally (I)

E - Extravert / I - Introvert

2. How do you look at things? Big Picture (N) or Detail (S)

N - Intuitive / S - Sensing

3. How do you approach tasks? Seek Results (T) or Seek Harmony (F)

T - Thinking / F - Feeling

4. How organised are you? Write lists (J) or Go with the flow (P)

J - Judging / P - Perceiving

Scarily spot on or way off the mark? Let us know what you think of your profile @RobsonDob #SakuraBook

INDEX OF PROFILES

ENFJ

THE GIVER

DESCRIPTION

People persons with excellent people skills. They give love and support. They focus on understanding and encouraging others. They make things happen for people and gain great satisfaction from this. They can make people do what they want of them. Their motives are usually unselfish.

They are so outwardly focussed they may need to spend more time alone. This can be quite dark for them as they can be very hard on themselves. They focus on helping others achieve their own goals sometimes at the expense of their own. Although extraverted they are less likely to expose themselves than other extraverts. They have strong held beliefs but may hold back in expressing them. They can sometimes be lonely in the company of others, not revealing their true selves.

They do well in positions where they work with people, such as teaching and are great candidates for a social committee. They like planning more than achievement. They like a future of possibilities and can get easily bored with the present.

They have a strong need for close intimate relationships and will put in a lot of effort into cultivating these. Loyal and trustworthy. If they haven't found their place in the world they could be sensitive to criticism, worry excessively and feel guilty. Generally, they are warm, charming and gracious who see the good in others and help them succeed. They may need to be aware of their own needs.

CAREERS

Teacher, Psychologist, Consultant, Social Worker, Human Resources, Politician.

IDEAL MATCH

INFP - THE IDEALIST

ISFP - THE ARTIST

FAMOUS PEOPLE

Martin Luther King Jr, Nelson Mandela, Sheryl Sandberg, Oprah Winfrey, Morgan Freeman, Bono, Mikhail Gorbachev, Tony Blair, Matthieu Ricard, Joseph Goebbels, Nigella Lawson, Reese Witherspoon, Kate Winslet, Helena Bonham Carter, Jennifer Lawrence, Emma Stone, Bradley Cooper, Dakota Fanning, Anna Sophia Robb, Matt LeBlanc, Johnny Depp.

ENFP

THE INSPIRER (BEEF)

DESCRIPTION

Warm, enthusiastic, bright and full of potential. Live in a world of possibilities and become very passionate about things. They motivate and inspire more than others and can talk themselves into or out of just about anything. They love life and want to make the most of it.

With a broad range of skills, they are project oriented. May go through several careers in a lifetime. To others, they may seem directionless but they are actually very consistent. Live life in accordance with their strong set of values. Feel the need to live as their true self. They are intense with a highly evolved sense of self and developed values. They need to focus on following through with their projects as this can be a problem.

Unlike other extraverts, they need to spend time alone to centre themselves in accordance with their values. Most have great people skills and place great importance on their relationships. They have a strong need to be liked. Once comfortable with themselves they excel in bringing out the best in others. They have an intuitive ability to understand others quickly and can relate to others on their level.

They see the world in exciting ideas but detest the drudgery of everyday life. Place little importance on day to day maintenance tasks. ENFPs get bored easily. May find it difficult maintaining marital relations. Always seeing the possibility in what could be, may get bored with what actually is. They like excitement in their lives and should seek a partner who is comfortable with change and new experiences.

Work best in situations with people, flexibility and ideas. Can be productive with little supervision. Do well when they go into businesses by themselves. Charming, ingenious risk takers. Sensitive and people oriented.

CAREERS

Consultant, Entrepreneur, Writer, Journalist, Artist, Actor.

IDEAL MATCH

INTJ - THE SCIENTIST

INFJ - THE NURTURER

FAMOUS PEOPLE

Oscar Wilde, Arianna Huffington, Walt Disney, Alan Watts, Osho, Che Guevara, Fidel Castro, Hunter S. Thompson, Mark Twain, Aldous Huxley, Julian Assange, Anne Frank, Naomi Klein, Col. Gadafi, Hugo Chavez, Salvador Dali, Orson Welles, Oliver Stone, Ellen Degeneres, Robin Williams, Jennifer Aniston, Sandra Bullock, Alicia Silverstone, Gwen Stefani, Daniel Radcliffe, Sharon Stone, Hans Zimmer.

ENTP

THE VISIONARY (ROBSON DOB)

DESCRIPTION

Their primary interest is understanding the world around them. Absorbing ideas, processing information, they are extremely quick at sizing up a situation. They generally understand things quickly and with great depth. They tend to be flexible and adapt well to a wide range of tasks. Good at most things that interest them.

Can see possibilities everywhere and get enthusiastic and passionate about their ideas. They can spread their enthusiasm to others. They are much less interested in developing action plans to implement the ideas. Usually bouncing on to the next idea instead. They have a danger of not finishing what they start. They need to be careful to think ideas through and implement them.

Fluent conversationalists. Mentally quick. Enjoy verbal sparring. Love to debate issues and may switch sides during an argument or play devil's advocate. If the Feeling side is not developed they may not value others' input or may be perceived as harsh or insensitive. Under stress, they become overly obsessed with minor details which are usually inconsequential in the grand scheme of things.

Overall upbeat visionaries, they highly value knowledge and spend their lives seeking a greater understanding. Living in a world of possibilities they get excited by concepts, challenges and difficulties. Good problem solvers, creative, curious and clever.

CAREERS

Lawyer, Sales, Marketing, Entrepreneur, Psychologist, Actor, Engineer, Consultant, DJ.

IDEAL MATCH

INFJ - THE PROTECTOR

INTJ - THE SCIENTIST

FAMOUS PEOPLE

Socrates, Benjamin Franklin, Leonardo Da Vinci, Bertrand Russell, Barack Obama, Martin Scorcese, Hugh Grant, Catherine the Great, Niccolo Machiavelli, Steve Wozniak, Henry Kissinger, Mao Zedong, Jeremy Clarkson, Robert Downey Jr, John Cleese, Rowan Atkinson, Federico Fellini, Karl Lagerfeld, Stephen Fry, Bill Hicks, Gillian Anderson, Elizabeth Olsen, Salma Hayek, Claire Danes, Matthew Perry.

ENTJ

THE EXECUTIVE (ELIZA)

DESCRIPTION

Natural born leaders. Live in a world of possibilities where challenges need to be surmounted. They want to be the ones to surmount them. They absorb a large amount of information and make quick and decisive judgements. They like to take charge. Long-range thinkers. Come up with plans to turn problems around. Tireless hard workers.

They dislike mistakes being made twice and have no time for inefficiency. They can be harsh when their patience is tried and may not be naturally in tune with other people's feelings. Have difficulty in seeing things from outside of their own perspective. Do not have much patience with people who don't see things their way. Can be a forceful, intimidating and overbearing person if not monitored. Has a tremendous amount of personal presence which helps them to achieve their goals. Sometimes this can lead to alienation and self-aggrandisement. Make decisions quickly and verbalise their opinions.

Enjoy a powerful conversation and appreciate someone who stands up for themselves and their opinions. Not many do as the ENTJ can be very forceful. Tend to have beautiful, well furnished, efficiently run homes. Best paired with someone who has a strong self-image. ENTJs can be very focussed on their careers and may be absent from homelife. Assertive, innovative long-range thinkers. Translates theories into action plans. Forceful and achievers.

CAREERS

Business Executive, Lawyer, Judge, Entrepreneur.

IDEAL MATCH

INTP - THE THINKER

ISTP - THE MECHANIC

FAMOUS PEOPLE

Napoleon Bonaparte, Julius Caesar, Aristotle, Bill Gates, George Clooney, Katherine Hepburn, Jack Welch, Garry Kasparov, Aung San Suu Kyi, Joseph Stalin, Michael Douglas, Adele, Charlize Theron, David Letterman, Tea Leoni, Penn (of '& Teller' fame).

ESFJ

THE CONNECTOR (MR SAKURA)

DESCRIPTION

People persons. They love people. They take on board many details about people and want to like them. They have a special skill in bringing out the best in people. They are good at reading others and understanding their point of view. They want to be liked and things to be pleasant so are supportive of others. They make people feel good about themselves and so many people want to be around them.

They take their responsibilities very seriously and are very dependable. They value security and stability. They have a strong focus on detail. They tend to see what needs to be done long before others do and take personal pride in getting them done.

Warm and energetic, they need approval from others to feel good about themselves. They have a strong need to be liked and in control. They have well-formed opinions on how things should be. They will express them but they will also abide by the moral code of a group as opposed to any internal value system.

They are kind and generous. They would give you the shirt off their back if they thought it was useful. Their selfless quality is genuine and sincere. They have a natural tendency to want to control their environment.

Enjoy order and structure. ESFJs at their best are warm, sympathetic, helpful, cooperative, tactful, down-to-earth, practical, thorough, consistent, organised, enthusiastic and energetic. They enjoy tradition and security, and will seek stable lives that are rich in contact with friends and family.

CAREERS

Nurse, Teacher, Childcare, Physio, Doctor, Clergy, Social Worker, Mayor, Bar Owner, T-Shirt Designer, Revolutionary.

IDEAL MATCH

ISFP - THE ARTIST

INFP - THE IDEALIST

FAMOUS PEOPLE

Harry Truman, Larry King, Sarah Jessica Parker, Prince William, Elton John, Hugh Jackman, Whitney Houston, Penelope Cruz, Jennifer Garner, Mariah Carey, Celine Dion, Shania Twain, Ed Sheeran, Alicia Keys, Tyra Banks, Victoria Beckham, Selena Gomez, Ariana Grande, Idris Elba.

ESFP

THE PERFORMER

DESCRIPTION

Lively and fun, they enjoy being the centre of attention. They live in the here and now and love drama and excitement in their lives. With strong interpersonal skills, they are often the peacemaker as they are sympathetic and care about people's well-being.

Generous and warm they are usually very observant and will sense what is wrong with someone before others would. They are not great at planning but they are great at giving practical care. Fun loving, spontaneous and optimistic. There could be in danger of being over indulgent in the here and now without considering the consequences of their actions.

The world is their stage. They love to perform and put on a show. They would love nothing more than life to be one continual party where they are the host. Upbeat and enthusiastic, they get on with just about everyone. Once crossed, however, they will hold a grudge. Under stress, they can be overly negative.

Practical, although they hate structure and routine. They go with the flow, trusting their ability to improvise. They are hands on experience learners rather than textbook theorists. They like and understand aesthetic beauty. Are likely to have a large number of beautiful possessions and an artistically decorated home. They usually like the finer things such as good food and wine. They have strong bonds with animals and children.

The ESFP has a tremendous love for life. They like to bring others along and are fun to be around. They're flexible, adaptable, genuinely interested in people and usually kind-hearted. They have a special ability to get a lot of fun out of life, but they need to watch out for the pitfalls associated with living entirely in the moment.

CAREERS

Artist, Actor, Sales, Fashion Design, Interior Decorator, Photographer.

IDEAL MATCH

ISTJ - THE DUTY FULFILLER

ISFJ - THE NURTURER

FAMOUS PEOPLE

Bill Clinton, Richard Branson, Larry Ellison, Deepak Chopra, Paulo Coelho, Hugh Hefner, Quentin Tarantino, Steven Spielberg, Beyonce, Will Smith, Jamie Oliver, Robbie Williams, Nicky Minaj, Pink, Leonardo Di Caprio, Michelangelo, Howard Shultz, Tony Robbins, Benito Mussolini, Horatio Nelson, Denzel Washington, Katy Perry, Cameron Diaz, Harry Styles, Justin Bieber, Lindsay Lohan, Peter the Great, Brad Pitt.

ESTP

THE DOER (ADRIAN ROBSON)

DESCRIPTION

Straight shooting, the ESTP is a doer. They thrive on action. Blunt, straightforward risk takers, they get their hands dirty and get things done. They live in the here and now and place little importance on theory. They look at the facts, decide what needs to be done, take action and move onto the next thing.

They pick up on people's motivations and attitudes more than the other personality types; things like facial expressions and posture. Typically a couple of steps ahead. They see rules as guidelines and do what they have to do to get things done. They have their own sense of principals that they abide by. They won't be swayed to do anything they think is wrong.

They have flair, talk fast and live faster. They could be a gambler or a spendthrift. Story tellers and improvisers. They make it up as they go along and don't follow a plan. Without realising it they can sometimes be hurtful as they don't realise the impact of their actions and words.

They make decisions based on logic and fact. Not very intuitive, they don't trust their instincts or people who trust theirs. They may be exceptionally clever but they will not excel at school and higher education where theory is important. They need a career that is constantly fluid. They despise routine.

Naturally energetic and enthusiastic, they can sell anyone on any idea. They are masters at getting things started although they may not always follow through. ESTPs are practical, observant, fun-loving, spontaneous risk-takers with an excellent ability to quickly improvise an innovative solution. They're enthusiastic and fun to be with, and are great motivators. If an ESTP recognises their talents they can accomplish exciting things.

CAREERS

Facilitator, Sales, Police, Paramedic, Technical Support, Athlete, Wanderer round Asia with no apparent cause or direction.

IDEAL MATCH

ISFJ - THE NURTURER

ISTJ - THE DUTY FULFILLER

FAMOUS PEOPLE

Alexander the Great, Winston Churchill, Franklin D. Roosevelt, JFK, Donald Trump, Ernst Hemingway, Arthur Conan Doyle, Dale Carnegie, George W. Bush, Malcolm X, Thomas Edison, Hermann Goering, Al Capone, L. Ron Hubbard, Alfred Hitchock, Harry Houdini, Angelina Jolie, Meryl Streep, Madonna, Taylor Swift, Amy Winehouse, Judi Dench, Helen Mirren, Miley Cyrus, Megan Fox, Mila Kunis, Kevin Spacey.

ESTJ

THE GUARDIAN (HGL)

DESCRIPTION

They live in a world of facts and concrete needs. They live in the present and are constantly scanning their environment to make sure everything is running smoothly and systematically. They honour traditions and laws and have a fixed set of internal beliefs. They have no time for those who don't stick to the same set of values.

They like competence and efficiency. Take charge people, they have a natural affinity for leadership and have a clear view on how things should be. They are stringent in their planning and developing action plans to get things done. They speak their mind but can often come across as being bolshy or blunt. They are straight talking and honest, and are not afraid to give their opinion.

A model citizen and pillar of the community, they take their commitments seriously. Dependable and reliable. They can be fun and boisterous at social events. They can at times be rigid and overly detail oriented. Because they have such high self belief they may not appreciate others' input and opinions and may end up hurting others' feelings. They may apply logic and rationale in situations that require a level of emotional sensitivity.

When under stress they may feel undervalued or misunderstood, or that their efforts are being taken for granted. ESTJs value security and social order and feel obliged to promote these goals. They will mow the lawn, vote, join the PTA, attend home owners association meetings and generally do anything that they can to promote personal and social security. They put a lot of effort into everything that they do. Conscientious, practical, realistic and dependable.

CAREERS

Military Leader, Business Administrator, Manager, Police, Judge, Coach, Financial Officer, Teacher, Physiotherapist.

IDEAL MATCH

ISTP - THE MECHANIC

INTP - THE THINKER

FAMOUS PEOPLE

Henry Ford, Margaret Thatcher, Hilary Clinton, Theresa May, Condoleeza Rice, Michelle Obama, Tom Clancy, Alan Dershowitz, Saddam Hussein, Augusto Pinochet, Uma Thurman, Emma Watson, Courteney Cox, Alec Baldwin, Jenny McCarthy, Ivanka Trump.

INFJ

THE PROTECTOR (SALLY)

DESCRIPTION

Gentle, caring, complex and highly intuitive. Artistic and creative, they live in a world of hidden meanings and possibilities. It is the rarest of all the personality types with only 1% of the population being them.

Outwardly they focus on identifying the best system to getting things done. They are constantly defining and redefining their priorities in life. Internally they are very intuitive and know things without being able to pinpoint why and without detailed knowledge. Their intuitive side means that they have feeling about things and later find out that they are right. It is almost a psychic tendency which other personality types may scorn or scoff at.

As a result the INFJ is very protective of their inner self and may not openly share as much about themselves. They can be deep, complex, private individuals who from the outside can be difficult to understand. They hold back and are secretive. They trust in their instincts above all else. This could pose a problem if they are stubborn and don't listen to others' opinions.

They believe they are right. They can also be perfectionists who doubt themselves and don't think they are living up to their own potential. They believe in growth but don't necessarily take stock of their achievements.

Rarely at peace, they constantly think they should be doing something more.

Strong value systems. Live according to what they think is right. Have high expectations for themselves and their families. Natural nurturer. Patient, devoted and protective. They excel best in careers where they can be creative and independent. They are not great with detail oriented tasks. The INFJs do not necessarily make life easy for themselves but are capable of great depth of feeling and personal achievement.

CAREERS

Clergy, Medic, Dentist, Psychologist, Psychiatrist, Counselor, Social Worker, Musician, Artist, Photographer, Child Care, Journalist.

IDEAL MATCH

ENTP - THE VISIONARY

ENFP - THE INSPIRER

FAMOUS PEOPLE

Plato, Gandhi, Thomas Jefferson, Dante, Agatha Christie, Arthur Schopenhauer, Leo Tolstoy, Adolf Hitler, Ruhollah Khomeini, Osama Bin Laden, Leon Trotsky, Robert Mugabe, Marilyn Manson, Al Pacino, Edward Norton, Cate Blanchett, Michelle Pfeiffer, David Schwimmer, Derren Brown, George Harrison, Benedict Cumberbatch.

INFP

THE IDEALIST (GEORDIE SAM)

DESCRIPTION

They are focussed on making the world a better place. They look for meaning in their own life and finding out their purpose and how to best serve humanity. Perfectionists who drive themselves hard to achieve the goals they set.

Highly intuitive, they are on a constant mission to find out the truth. Every fact is sifted through their value system and then evaluated to see if it can help them refine their path. Thoughtful and considerate. Good listeners. Reserved from the outside, they have an intense passion for getting to know and understand people.

They do not like conflict and avoid it where possible. When they do, they focus less on who is right and wrong and more on how it makes them feel. They therefore make good mediators. To some they may appear irrational and illogical but that is because they have an understanding for different people's perspectives and feelings.

Usually flexible and laid-back they will be incited to passionately defend a cause if it is important to them. They identify causes, and although not detail oriented, will cover every detail when working towards one. They don't like hard facts and logic. They sometimes misuse it. In stress they might defend themselves by throwing out fact after fact which may not be correct in an emotional outburst.

High standards, they don't give themselves enough credit. Hard for them to work in a team because they set their standards so high. They like to be in control. May be awkward expressing self verbally but can be a talented writer. They need jobs where they can work towards some human good. Overall they can accomplish wonderful things for humanity and won't even take the credit.

CAREERS

Writer, Counsellor, Teacher, Psychologist, Psychiatrist, Musician, Religious Worker, Computer Programmer, Pilot.

IDEAL MATCH

ENFJ - THE GIVER

ESFJ - THE CONNECTOR

FAMOUS PEOPLE

Albert Camus, George Orwell, JRR Tolkien, CS Woolf, AA Milne, JK Rowling, Edgar Alan Poe, John Milton, William Blake, William Shakespeare, Vincent van Gogh, Hans Christian Anderson, John Lennon, Jim Morrison, Kurt Cobain, Ian Curtis, Tim Burton, Johnny Depp, Florence Welch, Thom Yorke, Jarvis Cocker, Morrissey, Andy Warhol, Jude Law, Heath Ledger, Chris Martin, Marlon Brando, Lisa Kudrow.

INTP

THE THINKER (ARMSTRONG)

DESCRIPTION

Live in a world of theoretical possibilities. They see things based on how they could be improved or what they could be turned into. Living in their own minds they analyse difficult problems, identify patterns and come up with logical explanations. They seek clarity.

Driven to build knowledge. Absent minded professors who value intelligence and apply logic. Less concerned with the outer world; they seek to solve problems to take society to a higher understanding. Their minds are active; constantly trying to find new theories or disprove old ones.

They may appear dreamy but that is only because they are active within their own head solving problems. They like to discuss their theories. When interacting with people they are very tolerant and flexible unless their firmly held beliefs have been compromised. They may be shy when around new people but gregarious and confident around those they know well.

They place a lot of importance on rational logical decision making. They have little time for emotionally subjective stances. This could lead them to be cold and out of tune with other's feelings. If their feeling side is not developed they can be cynical, critical and sarcastic. They may not exude the warmth required in some intimate relationships.

They despise routine and find paying bills and dressing appropriately a chore. They like ideas and facts. They express themselves in absolute truths. They may not be able to convey their ideas succinctly to others and are unlikely to adapt their style to improve the situation.

Independent, unconventional and original. They are not likely to place importance on traditional goals such as popularity or security. Complex characters, they can be restless and temperamental. Ingenious and unconventional in their thought patterns, they can be eccentric. They are the pioneers of thought.

CAREERS

Scientist, Physicist, Chemist, Strategic Planner, Mathematician, Professor, Engineer, Economist, Park Ranger, Tech Start Up/Artist.

IDEAL MATCH

ENTJ - THE EXECUTIVE

ESTJ - THE GUARDIAN

FAMOUS PEOPLE

Abraham Lincoln, Albert Einstein, Charles Darwin, Immanuel Kant, Marie Curie, Richard Dawkins, Adam Smith, Rene Descartes, Larry Page, Sergey Brinn, Jane Austen, John Le Carre, Sigourney Weaver, Alan Greenspan, Jesse Eisenberg.

INTJ

THE SCIENTIST

DESCRIPTION

Live in a world of ideas and strategic planning. They value intelligence, knowledge and competence. Have high standards and continuously strive to achieve. Tend to have high expectations of others. Observers of the world, they generate ideas and possibilities.

Quick to understand an idea, they are less interested in the theory but how it can be applied. They don't like to think for thinking's sake, instead they require action. They like to transform ideas into workable plans and solutions.

Natural born leaders but they usually find themselves in the background until they see a need to take over. They make good leaders as they observe their surroundings and see what needs to be fixed. They will find practical ways of doing that. They are the supreme strategists. They plan for every possible contingency. They are quick to make judgements and are decisive. Ambitious, self confident, long range thinkers.

They dislike messiness and inefficiency. They can come across as reserved. May not give credit or be overly affectionate. They don't see the need to express it. They may appear rigid from the outside but they are actually quite flexible and open to new ideas. Their focus is on solving problems in a practical way and are open to new patterns of thought.

When under stress may become overly obsessed with mindless, repetitive activity like over drinking. They may also become obsessed with detail that would not normally bother them. They need to remember to express themselves correctly as not everyone is as fast to have worked out their solutions. They have tremendous ability to accomplish their goals.

CAREERS

Scientist, Engineer, Teacher, Medic, Corporate Strategist, Business Manager, Organisational Leader, Computer Programmer.

IDEAL MATCH

ENFP - THE INSPIRER

ENTP - THE VISIONARY

FAMOUS PEOPLE

Isaac Newton, Karl Marx, Friedrich Nietzsche, Mark Zuckerberg, Elon Musk, Nikola Tesla, Stephen Hawking, John Maynard Keynes, Christopher Hitchens, Jean-Paul Sartre, Martin Luther, Vladimir Lenin, Francis Ford Coppola, James Cameron, Arnold, Schwarzenegger, Colin Firth, Jodie Foster, Roger Waters, Jay-Z.

ISFJ

THE NURTURER

DESCRIPTION

Warm hearted and kind, they look for the best in people. Value harmony and cooperation. Sensitive to others' feelings. Considerate and aware. Ability to bring out the best in people. They have a rich inner world not obvious to outside observers. Constantly take in information which they store for years. Not uncommon for them to remember a facial expression or conversation in precise detail.

They value security and kindness. Have a fixed idea of how things should be. Learn best by doing. Not interested in theory. Extremely dependable. Will carry out a task through to completion. Developed sense of space and function. Understands and highly in tune with the aesthetic appeal. Likely to have beautifully furnished homes. Good gift givers.

In tune with own feelings. Usually do not express themselves well. Can build up strong negative feelings towards people if no outlet is found. May turn into long held grudges which are difficult to change. They are unlikely to speak up when they are feeling down but will stand up for others. Strong sense of duty and responsibility. Reliable. Take their duties seriously. Have a difficult time saying no and so can be taken advantage of or become overburdened with too many tasks.

Need positive feedback even though this doesn't seem apparent. If not, can become discouraged and depressed. Under stress, harbour feelings of inadequacy and become convinced they can do no right. Generous and dependable. Strong ability to ensure things run smoothly. In touch with people. They need to remember not to be overly critical of themselves and give some of the warmth they apply to others to themselves.

CAREERS

Interior Decorator, Designer, Firefighter, Nurse, Office Manager, Social Worker, Paralegal, Priest, Shopkeeper, Accountant, Forest Ranger.

IDEAL MATCH

ESTP - THE DOER

ESFP - THE PERFORMER

FAMOUS PEOPLE

George A Custer, Jimmy Carter, George Bush Snr, Marcus Aurelius, Rosa Parks, Mother Teresa, Heinrich Himmler, Francisco Franco, Bashar al-Assad, Christopher Walken, Anthony Hopkins, Naomi Watts, Halle Berry, Dr Dre, 50 Cent, Kanye West, Gwyneth Paltrow, Katie Holmes, Prince Charles, Kate Middleton, Brian May, Tiger Woods, Bruce Willis.

ISFP

THE ARTIST

DESCRIPTION

Highly in tune with their senses, they are aware of the way things look, taste, smell, sound and feel. Strong aesthetic appreciation. Strong set of values. Need to live in accordance with what they feel is right. Will choose jobs that help them achieve these value-oriented goals.

Quiet and reserved. Difficult to get to know well. Hold back ideas and thoughts from people they don't know closely. Kind, gentle and sensitive. They may seem carefree and lighthearted but in reality they take life very seriously.

Taking in information and searching for meaning. Likely to be animal lovers and have an appreciation for nature. Original and independent, they need personal space. Appreciate people who get to know the real them. They are doers. Have difficulty with theory, need to see a practical application. Learn in a hands on environment. Think decision making needs to be subjective and not necessarily against some fixed laws.

Extremely perceptive and aware of how others feel. Usually very accurate in their assessments of people. Warm and sympathetic. Care for others. Will show their affection for people through actions rather than words. No desire to lead or control. No desire to be led or controlled.

They like time alone to assess facts against their value system and feel others should do the same. Don't give themselves enough credit for their achievements. Leads to an intense perfectionism and unnecessary harshness. Creates and appreciates art. Selflessly serves others. They can find life hard because they take it so seriously. They can make life rich and rewarding for themselves and those closest to them.

CAREERS

Artist, Musician, Designer, Child Care, Social Worker, Teacher, Psychologist, Vet, Forest Ranger, Paediatrician.

IDEAL MATCH

ESFJ - THE CONNECTOR

ENFJ - THE GIVER

FAMOUS PEOPLE

Jackie O, Johnny Ive, Thich Nhat Hanh, Nero, Rudolf Hess, Michael Jackson, Prince, Bob Dylan, Paul McCartney, Mick Jagger, Keith Richards, Jimi Hendrix, Kate Bush, David Bowie, Lady Gaga, Brad Pitt, Ryan Gosling, Audrey Hepburn, Elizabeth Taylor, Monica Bellucci, Nicole Kidman, Princess Diana, Marilyn Monroe, Britney Spears, Rihanna, Christina Aguilera, Pamela Anderson, John Travolta, David Beckham, Liv Tyler, Drew Barrymore, Prince Harry, Skrillex.

ISTP

THE MECHANIC (CLAIRE)

DESCRIPTION

Drive for understanding on how things work. Good at logical analysis. Strong reasoning skills. Not interested in concepts and theories unless they can see a practical application. Adventurous spirit. They are attracted to extreme sports. Thrive on action and are usually fearless. Independent. Don't follow rules. Like doing their own thing.

Become bored easily. Like spending time alone as this is when they can sort things out in their heads. Don't like planning or working at a desk all day. Adaptable and spontaneous. Strong technical skills. Make good technical leaders. Focus on detail and practical things. They can make quick decisions. Avoid making decisions based on personal judgements. Thinks they should be made impartially based on fact.

Not in touch with their feelings or those of others. Under stress may exhibit rash emotional outbursts or overshare their feelings (sometimes inappropriately). Excellent in times of crisis. Usually good athletes with good eye-hand coordination. Good at following through with a project. Thinking logically, they do well at school.

Majority of the time are patient with the exception of emotional outbursts when their patience is tried. Usually good at most things they do.

They are happiest in action oriented roles where logical analysis is required. Optimistic, loyal, full of good cheer, trusting and generous who don't want to be confined to routine.

CAREERS

Police Detective, Forensic Pathologist, Computer Programmer, Systems Analyst, Construction, Engineer, Mechanic, Farmer, Pilot, Entrepreneur, Accountant.

IDEAL MATCH

ESTJ - THE GUARDIAN

ENTJ - THE EXECUTIVE

FAMOUS PEOPLE

Steve Jobs, Jack Dorsey, Dalai Lama XIV, Donald Rumsfeld, Melania Trump, Vladimir Putin, Miles Davis, Frank Zappa, Stanley Kubrick, Clint Eastwood, Harrison Ford, Tom Cruise, Christian Bale, Eminem, Scarlett Johansson, Demi Moore, Snoop Dog, Phil Ivey, David Blaine, Woody Allen, Bill Murray, Bruce Lee, Daniel Craig, Simon Cowell, Karl Pilkington, Jenna Jameson.

ISTJ

THE DUTY FULFILLER (BAZ)

DESCRIPTION

Quiet and reserved individuals. Interested in security and peaceful living. Sense of duty. Leads them to fulfil task assigned. Organised and methodical. Can succeed at anything they undertake. Loyal, faithful, dependable. Place importance on honesty and integrity. Do good by their families and communities. Take things seriously. Can have an offbeat sense of humour. Believe in laws and traditions. Expect the same from others.

If they see a logical reason why things should be done differently they might adapt but generally tend to want to do things according to existing procedures and processes. May become overly obsessed with structure and doing things by the book. Dependable. Keeps promises. May have difficulty saying no and becoming overburdened. Will put a lot of energy in to practical tasks. If they see no need or logical point to something they will resist action. No use for theory, only practical applications.

Not naturally in tune with their own feelings or those of others. May have difficulty picking up on emotional needs. Being perfectionists themselves they take others' efforts for granted. They need to remember to pat themselves on the back. Excellent ability to take a task, define, scope and plan it and take it through to conclusion. Hard working. Do not allow obstacles to get in the way of their duties. Tend to have a strong sense of aesthetics and will have beautifully decorated, well maintained homes.

Under stress may go into 'catastrophe mode' where everything is wrong, berating themselves for everything they should have done or duties they failed to perform. They will have overly pessimistic visions of doom. They have plenty of potential. Logical, capable, disciplined. Highly effective at achieving their goals.

CAREERS

Business Executive, Administrator, Manager, Accountant, Financial Officer, Police Officer, Judge, Doctor, Dentist, Military Leader.

IDEAL MATCH

ESTP - THE DOER

ESFP - THE PERFORMER

FAMOUS PEOPLE

Duke of Wellington, George Washington, Dwight Eisenhower, Augustus, Warren Buffet, Jeff Bezos, Ingvar Kamprad (Ikea), Sigmund Freud, Richard Nixon, Woodrow Wilson, Angela Merkel, Robert De Niro, Sean Connery, Morgan Freeman, John Malkovich, Queen Elizabeth, Natalie Portman, Matt Damon.

ACKNOWLEDGEMENTS

I would like to thank Mr Sakura for his participation in this book. He has lived quite the life so far and I thank him for taking the time to tell me the story. Thank you for creating a revolution and a community in Vang Vieng and welcoming me to be a part of it.

Thank you to Joey and AH who have created the greatest party in the world, making Vang Vieng worth writing about. In the long term I hope the town will grow and prosper. I also love the work they have done supporting the EEFA program; teaching English to Lao children.

For more information, to donate or to volunteer please visit the websites:

www.justgiving.com/crowdfunding/robsondob

www.eefalaos.org

I would like to thank Sally for her flippant comment, 'I'm travelling for the next year, you should write his book,' which was the starting point of the whole thing.

Adrian Robson, for showing me that anyone can, in fact, write a book and his catchphrase from which the title of this book originates.

I would like to thank Beef, Eliza, Geordie Sam, HGL, Claire and Baz who made my time in Lao so memorable and gave me the encouragement to write.

Armstrong, for the adventure and especially for pushing me to complete this. He is a great friend who always motivates me to achieve and go beyond whatever I believe I am capable of. For that he's a pretty special friend. He is also the creative force behind the front and back covers which I am sure you will agree are quite striking. I hope these visuals will achieve cult-like status as they enmesh themselves in the fabric of society. Imagine a bus or plane load of people all reading this. Like when The Da Vinci Code first came out, you'd be on a plane and row after row of people would be reading it. I hope one day the buses in and out of Double V are like that. People will be asking, 'What's that pink book I keep seeing everywhere?'

The Lexpedition for putting me on to Lao as a destination in the first place.

My editor JB.

My proof readers: KH, KBG, NS, SR, JLW and NF.

HC for providing the brand artwork at the final hour.

SNVTVG for making it all come together on my return.

My parents for their love and support.

The website personalitypage.com which is a great resource for Myers-Briggs information.

Everyone at Rocket Angel Media who turned this into reality.

And to you, the reader, for taking the time to read and enjoy it. I hope you are inspired!

Thank you all!

ONE FINAL ENCORE

SECRET BONUS GAME

D R I N K T R I P L E

H O D O U B L E E E S

G N E I V G N A V T T

T A B L E A T U B E U

L O O P D B E O E P I

E S S A Y S D I H P F

S I N G L E N A N S A

Y A C T L A O P D R T

C O I N C I D E N C E

Find all words relating to Sakura. Unused letters reveal a hidden message.

Robson Dob went to Lao in search of relaxation, adventure and new experiences. He found all of that and a whole lot more; including a T-shirt, that led him to discover himself, his creativity and the greatest party on the planet. He is currently working on his second book, his music and the next big thing. He splits his time between Europe, Asia and the Americas.

TELL YOUR FRIENDS

Your Turn

Create more coincidences and maybe even fate. Give this book to someone or leave it in a random location. Don't think about it too much, just let your energy decide who or where. Fill in your details, take a photo and post to social media. We can track each copy as it travels the world! Ichigo Ichie.

Date	Social Media Contact	I found this book / I left this book
______	@__________________	________________/________________
______	@__________________	________________/________________
______	@__________________	________________/________________
______	@__________________	________________/________________
______	@__________________	________________/________________
______	@__________________	________________/________________
______	@__________________	________________/________________
______	@__________________	________________/________________
______	@__________________	________________/________________
______	@__________________	________________/________________
______	@__________________	________________/________________
______	@__________________	________________/________________
______	@__________________	________________/________________

Share a photo and be part of the story @RobsonDob #SakuraBook

Read this book for free? Why not make a contribution to EEFA
www.justgiving.com/crowdfunding/robsondob

Printed in Great Britain
by Amazon